# Antennas for Industrial and Medical Applications with Optimization Techniques for Wireless Communication

The text begins by covering the fundamental concepts and new advances in the field of antenna theory, antenna hardware, and propagation. It further explains the designing of metamaterials microstrip patch antennas for medical applications, photonic crystals of millimeter wave signals for 5G communications, dual-band miniaturized circular antennas for wireless networks, and ultra-thin compact flexible antennas for wearable applications.

This book:

- Presents the design and development of S-shaped and T-shaped microstrip path antennas for industrial applications.
- Highlights the use of W-shaped and metamaterials microstrip patch antennas for medical applications.
- Covers photonic crystals of millimeter wave signals for 5G communications.
- Showcases the importance of compact and wideband slot antenna for wireless communications.
- Illustrates the design of an ultra-thin compact flexible antenna for wearable applications.

It is primarily written for senior undergraduates, graduate students, and academic researchers in the fields of electrical engineering, electronics and communications engineering, antenna design, and microwave engineering.

# Antennas for Industrial and Medical Applications with Optimization Techniques for Wireless Communication

Edited by
S. Kannadhasan and R. Nagarajan

CRC Press is an imprint of the
Taylor & Francis Group, an **informa** business

MATLAB® is a trademark of The MathWorks, Inc. and is used with permission. The MathWorks does not warrant the accuracy of the text or exercises in this book. This book's use or discussion of MATLAB® software or related products does not constitute endorsement or sponsorship by The MathWorks of a particular pedagogical approach or particular use of the MATLAB® software.

Designed cover image: Shutterstock

First edition published 2025
by CRC Press
2385 NW Executive Center Drive, Suite 320, Boca Raton FL 33431

and by CRC Press
4 Park Square, Milton Park, Abingdon, Oxon, OX14 4RN

*CRC Press is an imprint of Taylor & Francis Group, LLC*

© 2025 selection and editorial matter, S. Kannadhasan and R. Nagarajan; individual chapters, the contributors

Reasonable efforts have been made to publish reliable data and information, but the author and publisher cannot assume responsibility for the validity of all materials or the consequences of their use. The authors and publishers have attempted to trace the copyright holders of all material reproduced in this publication and apologize to copyright holders if permission to publish in this form has not been obtained. If any copyright material has not been acknowledged please write and let us know so we may rectify in any future reprint.

Except as permitted under U.S. Copyright Law, no part of this book may be reprinted, reproduced, transmitted, or utilized in any form by any electronic, mechanical, or other means, now known or hereafter invented, including photocopying, microfilming, and recording, or in any information storage or retrieval system, without written permission from the publishers.

For permission to photocopy or use material electronically from this work, access www.copyright.com or contact the Copyright Clearance Center, Inc. (CCC), 222 Rosewood Drive, Danvers, MA 01923, 978-750-8400. For works that are not available on CCC please contact mpkbookspermissions@tandf.co.uk

*Trademark notice*: Product or corporate names may be trademarks or registered trademarks and are used only for identification and explanation without intent to infringe.

ISBN: 978-1-032-77497-8 (hbk)
ISBN: 978-1-032-90912-7 (pbk)
ISBN: 978-1-003-56048-7 (ebk)

DOI: 10.1201/9781003560487

Typeset in Sabon
by codeMantra

# Contents

# Preface

Microwave radio transmission is frequently used in point-to-point communication systems on the surface of the earth. Mostly, it can be used in satellite communications and in deep space radio communications. Additionally, microwave radio bands are used for radars, radio navigation systems, sensor systems, and radio astronomy. We plan to focus on advancements and innovations in microwave photonics and communication technology, including microwave, wireless, and optical communications for the exchange of information, new trends in antenna theory and techniques, antenna hardware, and propagation studies.

In recent years, antenna design has gotten a lot of attention. This is due to a surge in interest in a variety of applications, ranging from IoT to low-frequency long-range applications to high-frequency mmWave 5G mobile technology. Future phones and base stations will need multimode antenna technology that is both energy efficient and can function in the millimeter wave band in conjunction with legacy 4G and sub-6GHz 5G. Antennas should be small in size, but they must meet technical criteria such as greater power, wider bandwidth, higher gain, and insensitivity to human users' hand-held influence. This necessitates highly creative antenna design solutions that can function in both single and MIMO/Array configurations. Low-cost, intelligent antennas, and new radio propagation modeling and prediction tools for future wireless networks must arise to support the new frequency bands and wireless system topologies, among other difficulties in the antenna and propagation sectors. In this regard, possible solutions include mmWave beamforming, massive MIMO methods, and distributed antenna systems, which transform the current notion of wireless access networks (cellular structures with fixed base station locations) to a user- or even service-oriented perspective. 5G will enable the full potential of the Internet of Things by delivering much higher mobile broadband rates, reduced latency, and dependable communications (IoT). In addition to an improved broadband connection, this will enable new services such as tactile communications, smart manufacturing, and cities. The usage of the millimeter wave band, which will provide a network of tiny cells enabling high-capacity and area-efficient hotspot zones, is critical to 5G. The next 5G system will be a true mobile multimedia communication platform, including not just legacy heterogeneous mobile networks but also sophisticated radio interfaces and the ability to operate at millimeter wave frequencies to make use of the vast amount of available capacity. We welcome scholars and practicing engineers to submit original research papers that address antenna and optical antenna.

# Editors

**S. Kannadhasan** is working as an Associate Professor and HOD in the Department of Electronics and Communication Engineering at Study World College of Engineering, Coimbatore, Tamil Nadu, India. He completed his Ph.D. in the field of smart antennas at Anna University in 2022. He has 13 years of teaching and research experience. He obtained his B.E. in ECE from Sethu Institute of Technology, Kariapatti, in 2009 and his M.E. in Communication Systems from Velammal College of Engineering and Technology, Madurai, in 2013. He obtained his M.B.A. in Human Resources Management from Tamil Nadu Open University, Chennai. He has published around 110 papers in reputed international journals indexed by SCI, Scopus, Web of Science, and Major Indexing, and more than 250 papers have been presented or published in national and international journals and conferences. In addition, he has also contributed a book chapter. He also serves as a board member, reviewer, speaker, session chair, and member of the advisory and technical committees of various colleges and conferences. He also attends various workshops, seminars, conferences, faculty development programs, STTP, and online courses. His areas of interest are smart antennas, digital signal processing, wireless communication, wireless networks, embedded systems, network security, optical communication, microwave antennas, electromagnetic compatibility and interference, wireless sensor networks, digital image processing, satellite communication, cognitive radio design, and soft computing techniques. He is a member of SMIEEE, ACM, IET, ISTE, FIEI, FIETE, CSI, IAENG, SEEE, IEAE, INSC, IARDO, ISRPM, IACSIT, ICSES, SPG, SDIWC, IJSPR, and the EAI Community.

**R. Nagarajan** received his B.E. in Electrical and Electronics Engineering from Madurai Kamarajar University, Madurai, India, in 1997. He received his M.E. in Power Electronics and Drives from Anna University, Chennai, India, in 2008. He received his Ph.D. in Electrical Engineering from Anna University, Chennai, India, in 2014. He has worked in the industry as an

Electrical Engineer. He is currently working as a Professor of Electrical and Electronics Engineering at Gnanamani College of Technology, Namakkal, Tamil Nadu, India. His current research interest includes power electronics, power system, soft computing techniques, and renewable energy sources.

# Contributors

**Shakil Ahmed**
Department of Electronics and Communication Engineering
Jalpaiguri Government Engineering College
Jalpaiguri, India

**M. Aishwarya**
Sri Sairam Institute of Technology
Chennai, India

**A. Arunraja**
Department of Electronics and Communication Engineering
Christ (Deemed to be University)
Bangalore, India

**T. N. Avushatha**
Department of Electronics and Communication Engineering
Kamaraj College of Engineering and Technology
Virudhunagar, India

**Sheetal Bukkawar**
Department of Information Technology
Saraswati College of Engineering
Navi Mumbai, India

**R. Chandru**
Department of Electronics and Communication Engineering
Sri Ramakrishna Engineering College
Coimbatore, India

**Manjusha Deshmukh**
Department of Computer Science Engineering
Saraswati College of Engineering
Navi Mumbai, India

**S. Allwin Devaraj**
Department of Electronics and Communication Engineering
Francis Xavier Engineering College
Tirunelveli, India

**V. Karthiga Devi**
Kamaraj College of Engineering and Technology
Virudhunagar, India

**S. Dhanyasree**
KIT-Kalaignar Karunanidhi Institute of Technology
Coimbatore, India

**Joohi Garg**
Department of ECE
MNIT
Jaipur, India

**K. B. Gurumoorthy**
Department of Electronics and Communication Engineering
KPR Institute of Engineering and Technology
Coimbatore, India

**B. Hemalatha**
Sri Sairam Institute of Technology
Chennai, India

**Vaibhav S. Hendre**
Department of Electronics and Telecommunication
G H Raisoni College of Engineering and Management
Pune, India
and
Savitribai Phule Pune University
Pune, India

**S. Jegadesh**
KIT-Kalaignar Karunanidhi Institute of Technology
Coimbatore, India

**Mandar P. Joshi**
Department of Electronics and Telecommunication
GES's R H Sapat College of Engineering, MS & R
Nashik, India
and
Savitribai Phule Pune University
Pune, India

**Kailash V. Karad**
Department of Electronics and Telecommunication
GES's R H Sapat College of Engineering, MS & R
Nashik, India
and
Savitribai Phule Pune University
Pune, India

**N. Kavitha**
Department of Electronics and Instrumentation Engineering
Hindusthan College of Engineering and Technology
Coimbatore, India

**M. Gopi Krishnan**
KIT-Kalaignar Karunanidhi Institute of Technology
Coimbatore, India

**A. Kishore Kumar**
Department of Robotics & Automation
Sri Ramakrishna Engineering College
Coimbatore, India

**Sudip Mandal**
Department of Electronics and Communication Engineering
Jalpaiguri Government Engineering College
Jalpaiguri, India

**R. Mathu Mietha**
Kamaraj College of Engineering and Technology
Virudhunagar, India

**T. Nivethitha**
Department of Electronics and Communication Engineering
Sri Eshwar College of Engineering
Coimbatore, India

**P. Palaniyammal**
Department of Electronics and Communication Engineering
Study World College of Engineering
Coimbatore, India

**M. Pandimadevi**
Sethu Institute of Technology
Virudhunagar, India

**M. Parisa Beham**
Sethu Institute of Technology
Virudhunagar, India

**P. K. Poonguzhali**
Department of Electronics and Communication Engineering
Hindusthan College of Engineering and Technology
Coimbatore, India

**R. Prabha**
Sri Sairam Institute of Technology
Chennai, India

**M. Mary Adline Priya**
Saveetha Engineering College
Chennai, India

**Kabita Purkait**
Kalyani Government Engineering College
Kalyani, India

**S. Esakki Rajavel**
Department of Electronics and Communication Engineering
Karpagam Academy of Higher Education
Coimbatore, India

**Jaswantsing L. Rajput**
Department of Electronics and Telecommunication Ramrao Adik Institute of Technology
Navi Mumbai, India

**K. Ramasamy**
KIT-Kalaignar Karunanidhi Institute of Technology
Coimbatore, India

**A. Sadhana**
Sri Sairam Institute of Technology
Chennai, India

**Dipankar Saha**
Global Institute of Management & Technology
Krishnagar, India

**B. A. Sapna**
KIT-Kalaignar Karunanidhi Institute of Technology
Coimbatore, India

**M. M. Sharma**
Department of ECE
MNIT
Jaipur, India

**N. M. Mary Sindhuja**
Kamaraj College of Engineering and Technology
Virudhunagar, India

**V. Subashini**
Sri Sairam Institute of Technology
Chennai, India

**R. Tamilselvi**
Sethu Institute of Technology
Virudhunagar, India

**Cynthia Anbuselvi Thangaraj**
Department of Electronics and Communication Engineering
SEA College of Engineering and Technology
Bangalore, India

Chapter 1

# Microwave metamaterial absorbers

## Design and applications

*Joohi Garg and M. M. Sharma*

## 1.1 INTRODUCTION

Metamaterials are also known as artificial materials whose properties are not found in nature [1,2]. These materials have gained huge attention from researchers in the past decade due to their designing capability that originally does not exist. One of the uses of the metamaterial is to design sub-wavelength unit cells, also known as meta molecules or atoms, which exhibit novel electric and/or magnetic responses to incident electromagnetic (EM) waves [3–5]. A metamaterial absorber [6] is a type of metamaterial that has the capability to efficiently absorb EM waves. The development of metamaterials results in a series of intriguing applications, such as a cloak of invisibility [7,8], giant optical chirality [9,10], wave-front control [11,12], surface plasmon manipulations [13–15], as well as antennas of compact sizes and enhanced directionalities [16–18]. There is an inevitable intrinsic loss in the applications of metamaterials. Researchers have put a significant amount of effort to obtaining the low-loss devices through the application of restructuring the geometries [19–21]. Metamaterials are resonant in nature. Initially, metamaterial-based absorbers have very low bandwidth and are polarization sensitive, due to which they cannot be used for practical applications. One of the earliest reported RADAR absorbers is the Salisbury screen, where a single resistive sheet is placed at $\lambda/4$ distance from a conducting ground plane. However, its highly narrowband response limits its applications [22]. Extending the single sheet Salisbury absorber to the multilayered Jaumann absorber improves the bandwidth but on the cost of increased thickness and larger weight. Conventional structures such as square loops, hexagonal branches, Jerusalem crosses, trumpet-shaped structures and many more structures have been evaluated in the literature as having fractional bandwidths of 45%–65% of the center frequency [23]. Furthermore, the use of a high number of lumped resistors complicates production and elevates the entire cost of the structures.

According to K. N. Rozanov [24] there exists a limit for the maximum possible bandwidth achievable for a specific thickness. The thickness to bandwidth bound is given by

DOI: 10.1201/9781003560487-1

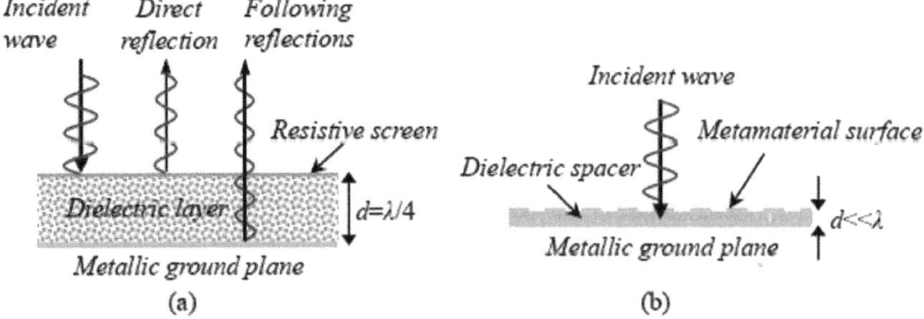


*Figure 1.1* (a) Salisbury screen and (b) perfect metamaterial absorber [39].

$$d \geq \frac{\left| \int_0^\infty \ln \left| R(\lambda) \right| d\lambda \right|}{2\pi^2}$$

where "$d$" is the absorber thickness, "$\lambda$" represents the wavelength and "$R$" is the reflectance.

Hence, achieving the widest bandwidth with the minimum possible total thickness is a challenging task [24]. Multiband absorptive metamaterial was produced by using several resonant unit cells and merging them in a single structure configuration. Whenever resonance frequencies are close to each other, broadband absorption is possible. Multilayered structures and various architectures also exhibit broadband absorption [25–27]. We may, however, use external biases to alter the absorptiveness and frequencies of absorbers by integrating active media as shown in Figure 1.1.

Currently, autonomously controlled wave absorbers are being aggressively pushed as wave absorbers based on innovative principles. Active metamaterial absorbers are also being used in many applications. Next, the electromagnetic absorber is thoroughly discussed from a variety of perspectives.

### 1.1.1 Theory of impedance matching

A metamaterial absorber is a sandwiched structure made up of an array of specific metallic patterns on one side of a substrate and a highly conductive metallic ground plane on the other. The electric permittivity and magnetic permeability of the metamaterial are $\epsilon = \epsilon_0 \epsilon_r(\omega)$ and $\mu = \mu_0 \mu_r(\omega)$, respectively, where $\epsilon_0$ and $\mu_0$ are the free space permittivity and permeability. $\epsilon_r$ and $\mu_r$ are frequency-dependent relative permittivity and relative permeability of the medium. Due to the presence of the ground plane, no signal can be transmitted through it, so absorptivity is solely dependent on the reflection coefficient [28–30].

$$A = 1 - |S_{11}|^2 - |S_{12}|^2$$

Since $S_{12}$=0, which implies

$$A = 1 - |S_{11}|^2$$

The reflectivity ($R$) of the metamaterial is determined by the Fresnel formula of reflection [29].

$$R_{\mathrm{TE}} = |r_{\mathrm{TE}}|^2 = \left| \frac{\mu_r \cos\theta - \sqrt{n^2 - \sin\theta}}{\mu_r \cos\theta + \sqrt{n^2 - \sin\theta}} \right|^2$$

$$R_{\mathrm{TM}} = |r_{\mathrm{TM}}|^2 = \left| \frac{\mu_r \cos\theta - \sqrt{n^2 - \sin\theta}}{\mu_r \cos\theta + \sqrt{n^2 - \sin\theta}} \right|^2$$

Transverse electric (TE) and transverse magnetic (TM) polarized waves are denoted by the subscripts TE and TM, respectively, and $n = \sqrt{\epsilon_r \, \mu_r}$ represents the metamaterial's effective refractive index. We have $\theta$=0° for the case of normal incidence, hence these equations simplify to [29]:

$$R = \left| \frac{Z - Z_0}{Z + Z_0} \right|^2 = \left| \frac{\sqrt{\mu_r} - \sqrt{\epsilon_r}}{\sqrt{\mu_r} + \sqrt{\epsilon_r}} \right|^2$$

With $Z = \sqrt{\mu/\epsilon}$ being the impedance of the metamaterial and $Z_0 = \sqrt{\mu_0/\epsilon_0}$ being the impedance of free space. As a result of the metallic ground's zero transmissivity, the absorptivity will be as follows [29]:

$$A = 1 - R = 1 - \left| \frac{Z - Z_0}{Z + Z_0} \right|^2 = 1 - \left| \frac{\sqrt{\mu_r} - \sqrt{\epsilon_r}}{\sqrt{\mu_r} + \sqrt{\epsilon_r}} \right|^2$$

For impedance matching, $Z = Z_0$ or $\mu_r = \epsilon_r$ is the crucial condition for attaining complete absorption. It's worth mentioning that simultaneous electric and magnetic resonances are necessary to accomplish impedance matching in a metamaterial absorber. In the single resonant metamaterial, the impedance of the metamaterial will be mismatched either in electric condition or in magnetic condition resulting in no ideal absorber.

### 1.1.2 Interference theory

In a metamaterial absorber, current in the front and rear metallic layers is in an anti-parallel direction, which induces magnetic resonance and forms a coupled system. The front layer, which is designed with certain metallic patterns, serves as a partial reflection surface, allowing the nonlinear reflection and transmission coefficients to be adjusted. The highly conductive ground plane, on the other hand, acts as a perfect reflector, providing the EM wave that reflects on it with a 180° phase delay [31].

## 1.2 CLASSIFICATION OF ABSORBERS [32]

EM wave absorbers are classified based on a variety of characteristics such as structure, material and frequency band of applications [33].

### 1.2.1 Classification by configuration forms

EM wave absorbers can be classified based on the "number of layers" that make up each absorber layer and the "appearance form" of the absorber structure. EM wave absorbers come in a variety of shapes and sizes.

#### *1.2.1.1 Classification based upon number of layers*

##### *1.2.1.1.1 Single-Layer Absorber*

A single-layer absorber is an absorber constructed from a single layer of material. A metal plate consisting of aluminum, copper or a similar material is usually affixed to the rear of the absorber. EM wave absorbers made of ferrite, carbonyl iron, and other similar materials are examples of this kind.

##### *1.2.1.1.2 Two-Layer Absorber*

A two-layer absorber is an absorber with distinct material constants in both layers. This kind of configuration is utilised to enhance a single-layer absorber's properties in order to achieve broadband absorption.

##### *1.2.1.1.3 Multilayered absorber*

A wave absorber with three or more layers is known as a multilayered wave absorber. The wideband features of the multilayered wave absorber are acquired by increasing the number of layers, and this type of absorber may be utilized, for example, in an anechoic chamber.

### 1.2.2 Classification by frequency characteristics [32]

A reflection coefficient below –20 dB is considered suitable for an EM wave absorber, and the figure of merit may be derived as $\Delta f/f_0$ if the bandwidth cut by the level of –20 dB values is considered to be $\Delta f$, as shown in Figure 1.2 with the center frequency $f_0$.

#### *1.2.2.1 Narrowband absorber*

A single-layer absorber having a percentage figure of merit around 10%–20% is known as a narrow band absorber, as shown in Figure 1.2a. This type of absorber is employed when a restricted frequency range is required, such as in radar applications.

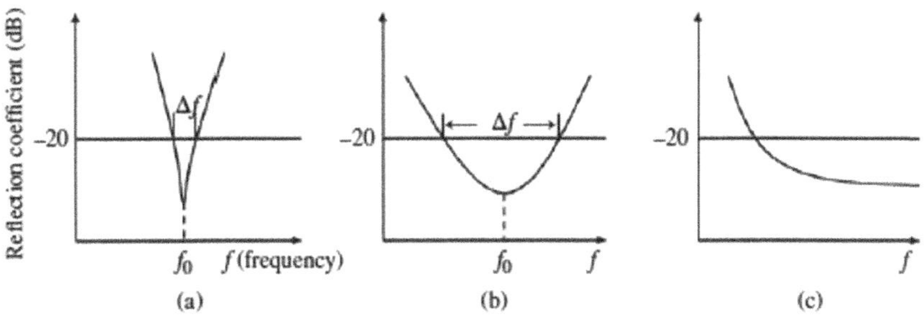


*Figure 1.2* Frequency band classification (a) Narrowband, $\Delta f/f_0 \times 100 = 10\%–20\%$, (b) Broadband, $\Delta f/f_0 \times 100 = 20\%–30\%$ and (c) Ultra-Wideband, $\Delta f/f_0 \times 100 = 30\%$ and above [32].

#### *1.2.2.2 Broadband absorber*

An absorber having a percentage figure of merit between 20% and 30% is known as a broadband absorber, as shown in Figure 1.2b.

#### *1.2.2.3 Ultra-wideband absorber*

An absorber having a percentage figure of merit greater than 30% is known as an ultra-wideband absorber, as shown in Figure 1.2c.

### 1.2.3 Classification by appearance

EM wave absorbers, which come in a variety of shapes and sizes, can be classified based on their appearance, as shown in Figure 1.3.

#### *1.2.3.1 Plane absorber*

A flat plate type wave absorber, as shown in Figure 1.3a, is made up of a structure with a flat surface against the incident EM wave direction. A ferrite wave absorber is a good example of this sort of wave absorber.

#### *1.2.3.2 Quarter wavelength absorber*

A quarter wave absorber is made by positioning a conductor plate a quarter wavelength distant from the film-shaped resistor, as shown in Figure 1.3b.

#### *1.2.3.3 Multilayered absorber*

Multilayered wave absorbers are built by stacking absorbent materials to get the matching characteristic by modifying each input impedance for each material progressively, as shown in Figure 1.3c.

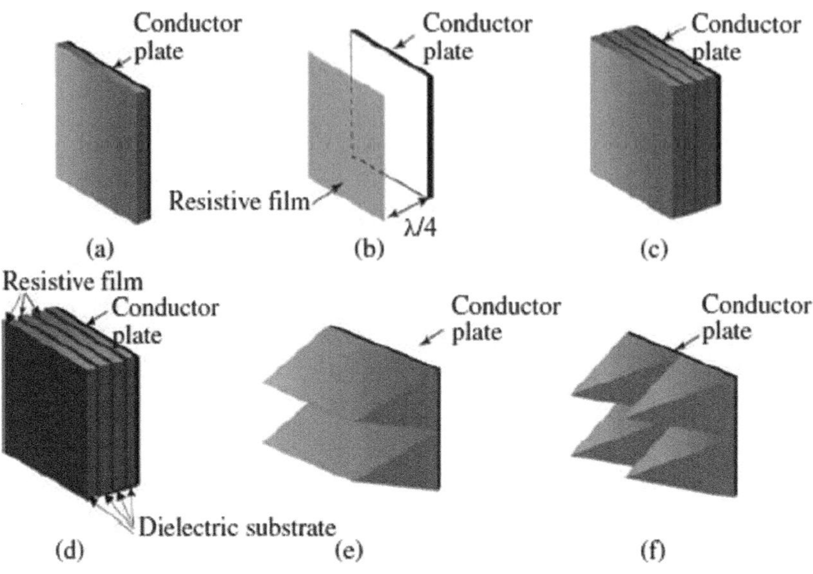


*Figure 1.3* Main wave absorber classification (a) Plane absorber, (b) Quarter wave absorber, (c) Multilayer absorber, (d) Jaumann absorber, (e) Saw-tooth absorber and (f) Pyramidal absorber [32].

#### *1.2.3.4 Jaumann absorber*

A Jaumann absorber is made up of stacked alternating resistive sheets and dielectric plates, as shown in Figure 1.3d.

#### *1.2.3.5 Saw-tooth shape absorber*

This type of EM wave absorber surface has a saw-toothed shape and has a tapering form, as shown in Figure 1.3e. The form of this absorber is also known as a Chevron absorber because of its layout. Although it is a single-polarization type of EM wave absorber, it is capable of efficiently absorbing EM waves throughout a large frequency range.

#### *1.2.3.6 Pyramidal wave absorber*

As the name suggests, the pyramidal wave absorber has a pyramidal structure on the EM wave incident side. It exhibits EM wave absorption characteristics for both polarized EM waves throughout a large frequency range, as shown in Figure 1.3f. We create this wave absorber by impregnating carbon. The substances are incorporated into urethane foam, Styrofoam, or other similar materials.

## 1.3 METAMATERIAL ABSORBERS BASED ON POLARIZATION INSENSITIVE FREQUENCY SELECTIVE SURFACES

### 1.3.1 Single-band polarization-insensitive metamaterial absorber [34]

In this section, a metamaterial-based absorber designed for a single frequency in the X-band range is discussed. The applications of this absorber are in satellite communication, aeronautical navigation, radio-determination, radiolocation in civil and radiolocation in military, etc. This structure is polarization-insensitive in nature for all incident angles.

This unit cell structure of the proposed metamaterial absorber is printed on an FR4 dielectric substrate which gives single-band characteristics. The dimensions of the unit cell are 4.1 mm×4.1 mm. In this unit cell, four square loop rings are alternatively connected, as shown in Figure 1.4a. A 6×6 array of the unit cell structure is shown in Figure 1.4b. The absorptivity vs frequency curve shows that the structure is resonant at 9.7 GHz with 99.48% absorptivity, as shown in Figure 1.4c. To determine the polarization-insensitive behavior of unit cell structure, measurements were taken at different polarization angles ranging from 0° to 60°, as shown in Figure 1.4d, confirming its polarization-insensitive behavior.

### 1.3.2 Dual-band polarization-insensitive metamaterial absorber [35]

In this section, a dual-band metamaterial absorber for X-band applications is discussed. This absorber is compact in size, very low thickness and potentially instructive for dual-band applications. The structure of thickness is only 0.6 mm which corresponds to $\lambda/53$ in terms of $\lambda$, wavelength corresponds to the lowest resonant frequency. The ultra-thin nature of the absorber makes it flexible and capable of applications in military applications, wireless communications, air surveillance radar and other X-band applications. The unit cell of the proposed absorber consists of two cross-ellipses surrounded by a U-shaped element printed on a grounded FR4 dielectric substrate, as shown in Figure 1.5a. Under normal incidence, two separate absorption peaks occur at 8.6 and 9.6 GHz, with absorptivities of 96% and 95.7%, respectively, as shown in Figure 1.5b.

The unit cell structure exhibits inductive behavior along the crossed ellipse and capacitance is provided by the U-shaped element and outer part of the ellipse. Ground and unit cell elements are separated by an FR4 dielectric substrate which also provides capacitance. Inductance and capacitance jointly form EM resonance, which produces dips in reflection coefficient and simultaneously shows maximization of absorption, as shown in Figure 1.5c.

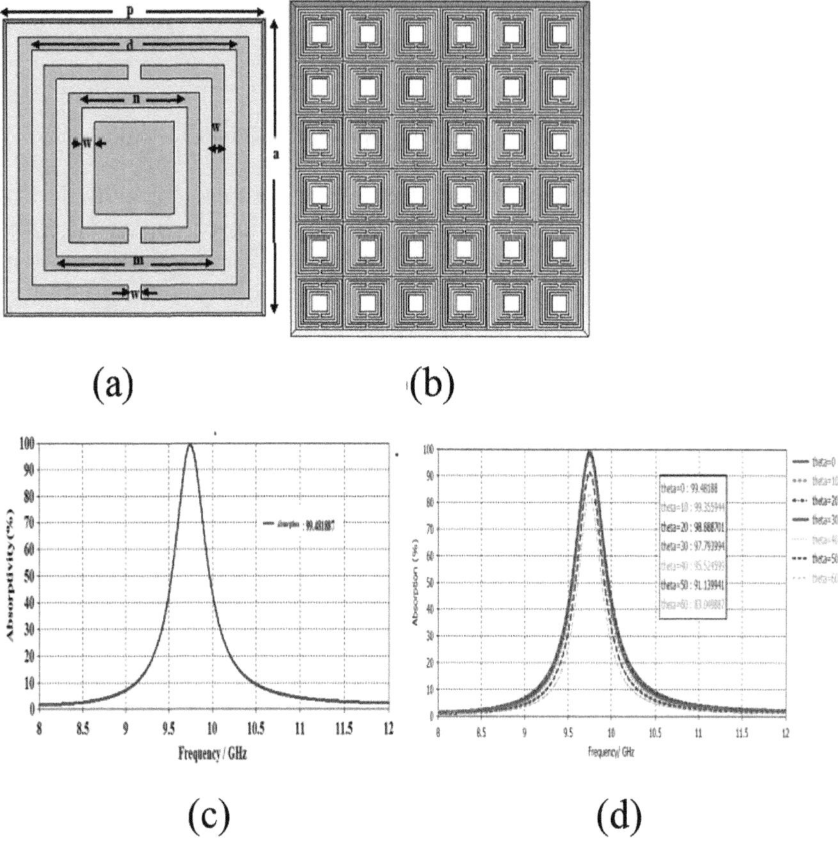


*Figure 1.4* (a) Proposed unit cell, (b) 6×6 array of proposed unit cell, (c) absorptivity vs frequency and (d) absorptivity of proposed structure at different incident angles [34].

The incident angles vary from 0° to 45° in TE mode, as shown in Figure 1.5d. It is evident that the differences are minimal and that the curves at any given angle almost completely overlap. The structure is polarization-insensitive due to the symmetrical design of the unit cell.

### 1.3.3 Broadband polarization-insensitive metamaterial absorber [36]

In this section, a broadband absorber for S-band applications is discussed. The structure incorporates circular and square rings that are printed on an FR4 dielectric substrate, as shown in Figure 1.6a. An array of 6×6 units is designed as shown in Figure 1.6b. This structure gives two absorption peaks, one at 3.6 GHz and the other at 3.8 GHz with peaks absorbance

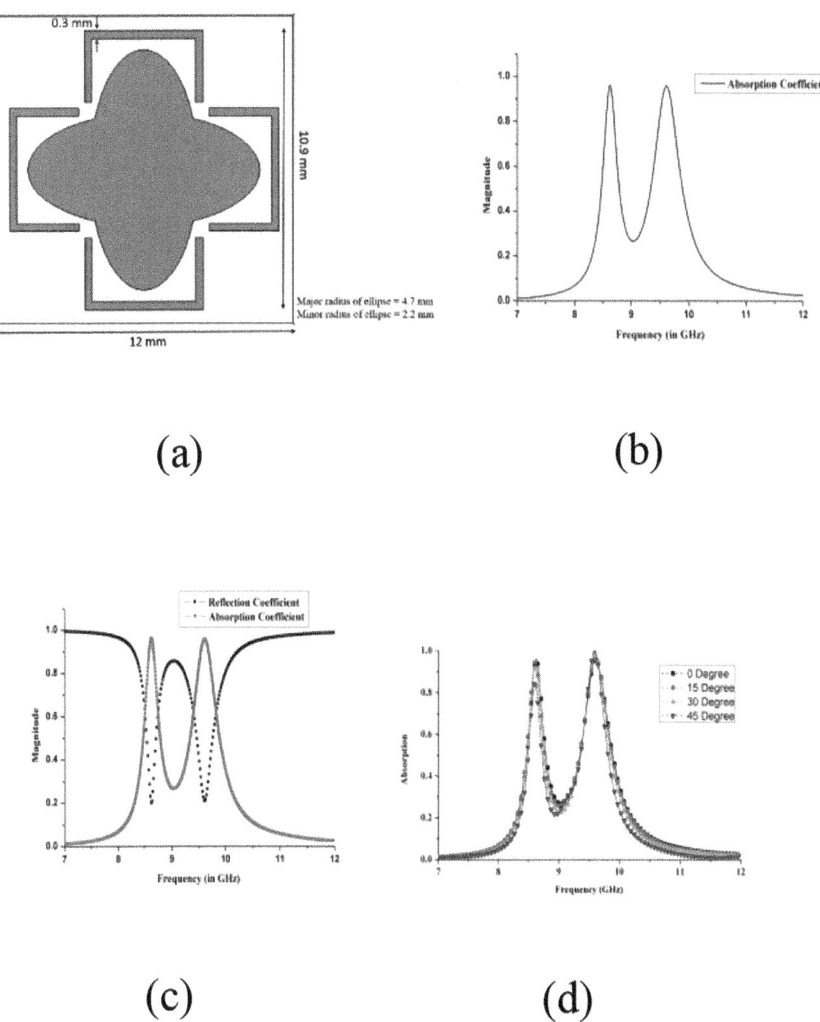


*Figure 1.5* (a) Front view of the unit cell of absorber, (b) Simulated absorption coefficient in TE mode, (c) Simulated absorption and reflection coefficient of absorber in TE mode, and (d) Variation of different incident angles theta from 0° to 45° [35].

of 97.26% and 97.03%, respectively. The two peaks are closely spaced so they get combined and form a broadband absorber, as shown in Figure 1.6c. It works in the S-band frequency range. To determine the polarization insensitive behavior of unit cell structure, it has been measured at different polarization angles ranging from 0° to 60° as shown in Figure 1.6d which shows very less variation while changing the incident angle. This absorber has many applications like FSS Earth stations, stealth technology used in the battlefield, and UWB applications.

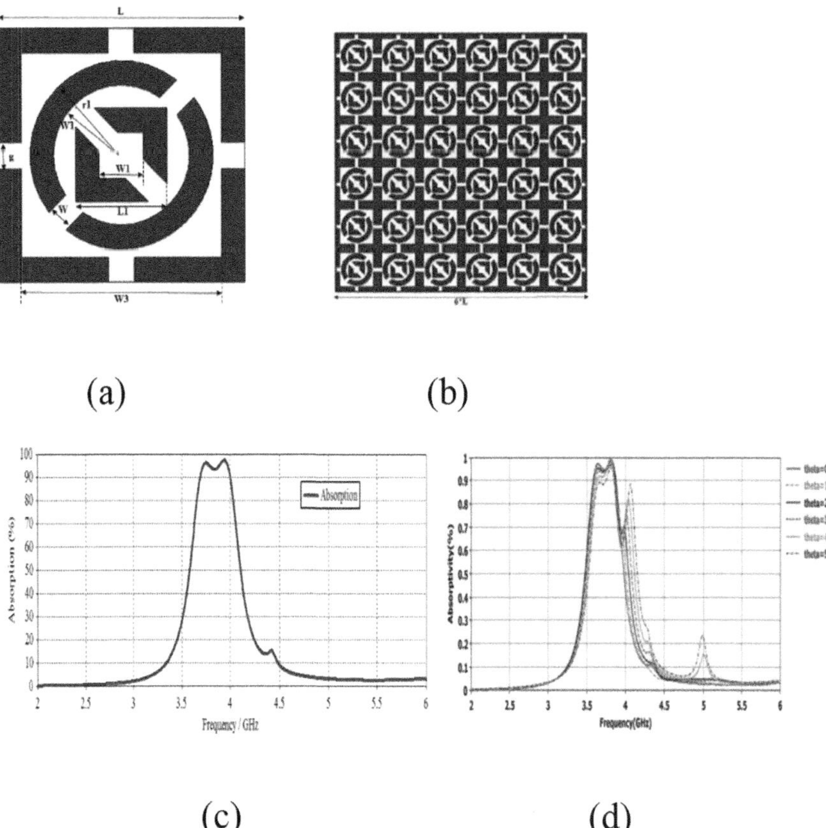


*Figure 1.6* (a) Front view of the unit cell of absorber, (b) 6×6 array of absorber, (c) Absorptivity vs frequency graph in TE mode, (d) Variation of different incident angles theta from 0° to 50° [36].

### 1.3.4 Multiband polarization-insensitive metamaterial absorber [37]

In this section, a quad-band polarization-independent metamaterial absorber for S-band, C-Band and X-band is discussed. The multiband absorber unit cell consists of meander line-type structure surrounded by a square ring printed on FR4 substrate backed by copper ground as shown in Figure 1.7a. The square ring is responsible for S-band, the meander line structure is responsible for C-band and the meander line with the plus structure is responsible for X-band absorption frequency. The absorber produces three diverse absorption peaks in TE mode and TM mode for frequencies 3.2, 6.25, 8.23 and 9.60 GHz with absorptivity of 97.7%, 97.6%, 93.1% and 99.6%, respectively, as shown in Figure 1.7b. The variation of theta and phi is carried out from 0° to 45° to observe their effect on absorption

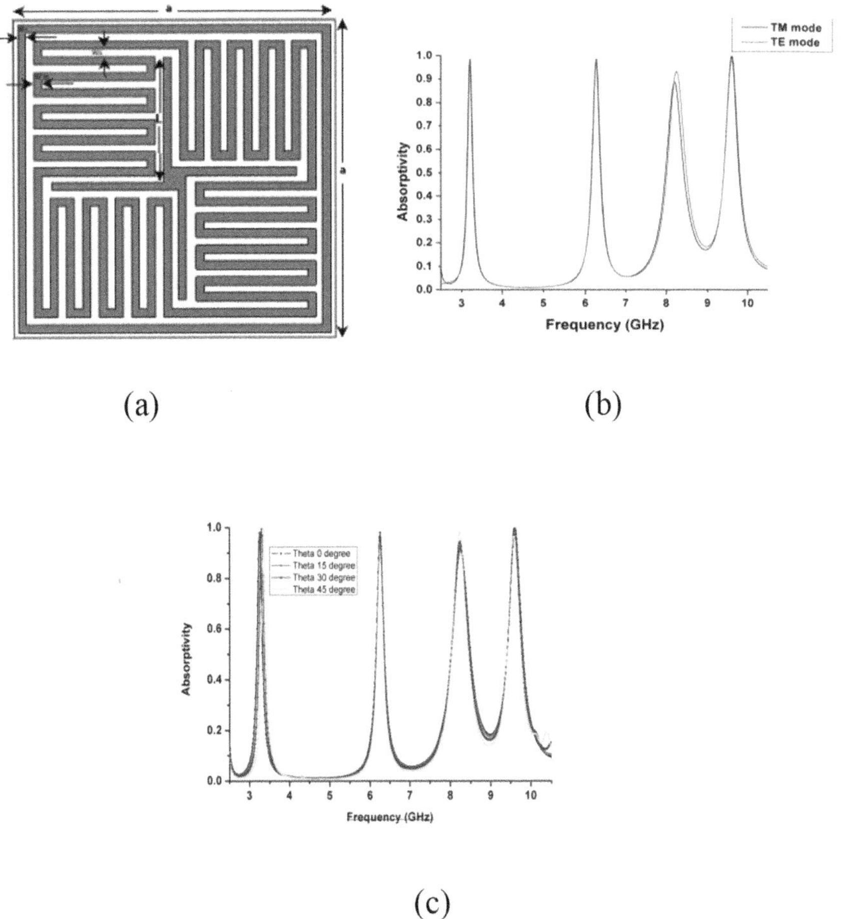


*Figure 1.7* (a) Front view of the unit cell of multiband absorber, (b) Absorptivity vs frequency graph in TE and TM mode, (c) Variation of different incident angles theta from 0° to 45° [37].

characteristics, as shown in Figure 1.7c, which shows that the absorber is polarization insensitive in nature. This absorber can be used for different applications like RCS reduction, military, security, radar applications EM interference, shielding and other applications.

### 1.3.5 Frequency selective surface-based terahertz absorber [38]

In this section, an FSS-based terahertz absorber is discussed. The structure of the absorber consists of a resonator, a metallic layer and a dielectric substrate, in which, the resonator and metallic layer are represented

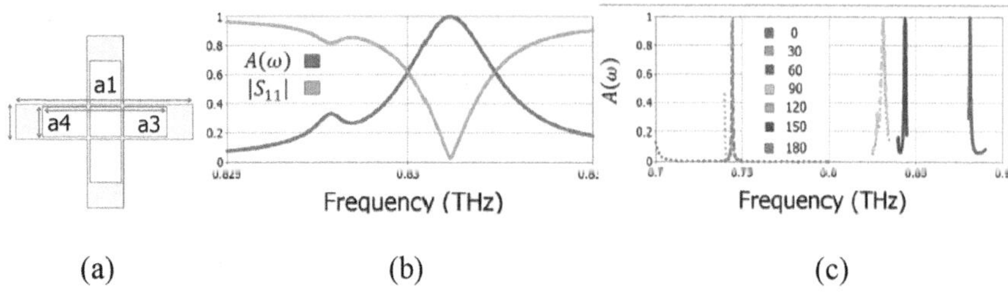


*Figure 1.8* (a) Front view of the unit cell of Terahertz absorber, (b) Absorptivity vs frequency graph, and (c) Variation of different incident angles from 0° to 180° [38].

by a silver sheet. The Quartz (Fused) dielectric substrate separates the resonator and metallic plate, which are aligned parallel to each other as shown in the Figure 1.8a. The absorptivity vs frequency curve shows the absorption at 0.83 THz with absorptivity of 99.92% as shown in Figure 1.8b. This absorber shows excellent absorption for all incidence angles. The incident angle is varied for this absorber.

The lowest absorption for 120° is approximately 0.83 THz, which is 97.14%, while the greatest absorption for 60° is around 0.88 THz, which is 99.98%, as shown in Figure 1.8b. As a result, changing the incidence angle just moves the resonance frequency, but it has no effect on the quality of the absorption.

## 1.4 DISCUSSION AND CONCLUSION

In this chapter, metamaterial-based different types of absorbers have been presented. The classification of metamaterial absorbers is based on frequency, appearance and number of layers. Some polarization-independent metamaterial absorber structures have been discussed. The structures are presented with the absorptivity and the effect of incident angle variation over the absorbers is also investigated. All of these structure shows complete absorption, and they can be readily customized for other frequencies by rescaling the structure.

According to the findings, regardless of polarization state or incidence angle, the metamaterial absorbers give complete absorption at resonant frequencies. The absorption value may be varied in various polarization states and incident angles. Within the frequency range, the presented metamaterial absorbers successfully depict single, dual, and even quadruple absorption bands in microwave and terahertz ranges. These absorbers have many benefits such as simplicity, flexibility, mechanical tunability, perfect absorption, etc. These metamaterial absorbers can be used for stealth, sensors, modulators, wireless communication, medical imaging and other applications. Using active components in absorbers can provide tunability, switchability as well as an increased absorption bandwidth. With the emergence of transparent electronics, like transparent thin film resistors, capacitors, inductors, diodes and transistors, active absorbers can be designed which are also optically transparent.

## REFERENCES


1. Caloz, C. and Ito, T., *Electromagnetic Metamaterials: Transmission Line Theory and Microwave Applications*, John Wiley & Sons, New York, 2006.
2. Marqués, R., Martín, F. and Sorolla, M., *Metamaterials with Negative Parameters: Theory, Design and Microwave Applications*, Wiley Interscience, New York, 2008
3. Zhang, X. and Liu, Z., Superlenses to overcome the diffraction limit, *Nature Materials*, 7, 435–441, 2008.
4. Zhu, W., Rukhlenko, I. D., Xiao, F. and Premaratne, M., Polarization conversion in U-shaped chiral metamaterial with four-fold symmetry breaking, *Journal of Applied Physics*, 115, 143101, 2014.
5. Schurig, D., Mock, J. J., Justice, B. J., Cummer, S. A., Pendry, J. B., Starr, A. F. and Smith, D. R., Metamaterial electromagnetic cloak at microwave frequencies, *Science*, 314(5801), 977–980, 2006.
6. Landy, N. I., Sajuyigbe, S., Mock, J. J., Smith, D. R. and Padilla, W. J., Perfect metamaterial absorber, *Physical Review Letters*, 100(20), 207402, 2008.
7. Schurig, D., Mock, J. J., Justice, B. J., Cummer, S. A., Pendry, J. B., Starr, A. F. and Smith D. R., Metamaterial electromagnetic cloak at microwave frequencies, *Science*, 314, 977–980, 2006.
8. Zhu, W., Shadrivov, I., Powell, D. and Kivshar, Y., Hiding in the corner, *Optics Express*, 19, 20827–20832, 2011.
9. Wang, B., Zhou, J., Koschny, T., Kafesaki, M. and Soukoulis, C. M., Chiral metamaterials: Simulations and experiments, *Journal of Optics A*, 11, 114003, 2009.
10. Zhu, W., Rukhlenko, I. D., Huang, Y., Wen, G. and Premaratne, M., Wideband giant optical activity and negligible circular dichroism of near-infrared chiral metamaterial based on a complementary twisted configuration, *Journal of Optics*, 15, 125101, 2013.
11. Kang, M., Wang, H. T. and Zhu, W., Wavefront manipulation with a dipolar metasurface under coherent control, *Journal of Applied Physics*, 122, 013105, 2017.
12. Cui, T. J., Qi, M. Q., Wan, X., Zhao, J. and Chen, Q., Coding metamaterials, digital metamaterials and programmable metamaterials, *Light: Science and Applications*, 3, e218, 2014.
13. Liu, Y., Zentgraf, T., Bartal, G. and Zhang, X., Transformational plasmon optics, *Nano Letters*, 10, 1991–1997, 2010.
14. Zhu, W., Rukhlenko, I. D. and Premaratne, M., Linear transformation optics for plasmonics, *Journal of the Optical Society of America B: Optical Physics*, 29, 2659–2664, 2012.

15. Zhu, W., Rukhlenko, I. D. and Premaratne, M., Maneuvering propagation of surface plasmon polaritons using complementary medium inserts, *IEEE Photonics Journal*, 4, 741–748, 2012.
16. Hong, W., Jiang, Z. H., Yu, C., Zhou, J., Chen, P., Yu, Z., Zhang, H., Yang, B., Pang, X., Jiang, M., Cheng, Y., Al-Nuaimi, M. K. T., Zhang, Y., Chen, J. and He, S., Multibeam antenna technologies for 5G wireless communications, *IEEE Transactions on Antennas and Propagation*, 65, 6231–6249, 2017.
17. Si, L. -M., Zhu, W. and Sun, H. -J., A compact, planar, and CPW-fed metamaterial-inspired dual-band antenna, *IEEE Antennas and Wireless Propagation Letters*, 12, 305–308, 2013.
18. Lin, F. H. and Chen, Z. N., Low-profile wideband metasurface antennas using characteristic mode analysis, *IEEE Transactions on Antennas and Propagation*, 65(4), pp. 1706–1713, 2017.
19. Cao, W., Singh, R., Al-Naib, I. A., He, M., Taylor, A. J. and Zhang, W., Low-loss ultra-high-Q dark mode plasmonic fano metamaterials, *Optics Letters*, 37, 3366–3368, 2012.
20. Zhu, W. and Zhao, X., Numerical study of low-loss cross left-handed metamaterials at visible frequency, *Chinese Physics Letters*, 26, 074212, 2009.
21. Zhu, W. and Zhao, X., Adjusting the resonant frequency and loss of dendritic left-handed metamaterials with fractal dimension, *Journal of Applied Physics*, 106, 093511, 2009.
22. Kong, L. B. Li, Z. W., Liu, L., Huang, R., Abshinova, M., Yang, Z. H., Tang, C. B., Tan, P. K., Deng, C. R. and Matitsine, S., Recent progress in some composite materials and structures for specific electromagnetic applications, *International Materials Reviews*, 58(4), 203–259, 2013.
23. Sheokand, H., Ghosh, S., Singh, G., Saikia, M., Srivastava, K. V., Ramkumar, J. and Anantha Ramakrishna, S., Transparent broadband metamaterial absorber based on resistive films, *Journal of Applied Physics*, vol. 122, 105105, 2017.
24. Rozanov, K. N., Ultimate thickness to bandwidth ratio of radar absorbers, *IEEE Transactions on Antennas and Propagation*, 48(8), 1230–1234, 2000.
25. Gu, S., Su, B. and Zhao, X., Planar isotropic broadband metamaterial absorber, *Journal of Applied Physics*, 104, 163702, 2013.
26. Cui, Y., Fung, K. H., Xu, J., Ma, H., Jin, Y., He, S. and Fang, N. X., Ultrabroadband light absorption by a sawtooth anisotropic metamaterial slab, *Nano Letters*, 12, 1443–1447, 2012.
27. Kim, Y. J., Yoo, Y. J., Kim, K. W., Rhee, J. Y., Kim, Y. H. and Lee, Y. P., Dual broadband metamaterial absorber, *Optics Express*, 23, 3861–3868, 2015.
28. Landy, N. I., Sajuyigbe, S., Mock, J. J., Smith, D. R. and Padilla, W. J., Perfect metamaterial absorber, *Physical Review Letters*, 100, 207402, 2008.
29. Zhu, W., Electromagnetic metamaterial absorbers: from narrowband to broadband, *Metamaterials and Metasurfaces*. IntechOpen, 2019. doi: 10.5772/intechopen.78581.
30. Mattiucci, N., Bloemer, M. J., Aközbek, N. and D'Aguanno, G., Impedance matched thin metamaterials make metals absorbing, *Scientific Reports*, 3, 3203, 2013.

31. Chen, H-T., Interference theory of metamaterial perfect absorbers, *Optics Express*, 20, 7165–7172, 2012.
32. Kotsuka, Y., *Electromagnetic Wave Absorbers*, Wiley, 2019.
33. Kotsuka, Y., *Fundamentals of Electromagnetic Wave Absorbers*, Wiley, 2019.
34. Singh, S. Yadav, N. and Chahar, R., Design and analysis of ultrathin polarization-insensitive metamaterial absorber for stealth technology applications, *2017 4th International Conference on Signal Processing and Integrated Networks (SPIN)*, Noida, pp. 193–195, 2017.
35. Garg, J., Yadav, S. and Sharma, M. M., A modified Jerusalem inspired bandpass FSS for multiband applications based on concentric ring slots, *Frequenz*, 2023, https://doi.org/10.1515/freq-2022-0218.2.
36. Garg, J., Yadav, S. and Sharma, M. M., A novel miniaturized loop based angularly stable and polarization independent multiband bandpass FSS structure for Wi-Max and WLAN applications, *Sādhanā*, 48, 14, 2023. https://doi.org/10.1007/s12046-022-02068-x.
37. Yadav, S., Abegaonkar, M. P., Sharma, M. M. and Jain, C. P., A quad-band polarization independent metamaterial absorber, *2017 IEEE Applied Electromagnetics Conference (AEMC)*, Guwahati, India, pp. 1–2, 2017.
38. Sabah, C., Dincer, F., Karaaslan, M., Unal, E. and Akwgol, O., Polarization-insensitive FSS-based perfect metamaterial absorbers for GHz and THz frequencies, *Radio Science*, 49(4), 306–314, 2014
39. Fante, R. L. and McCormack, M. T., Reflection properties of the Salisbury screen, *IEEE Transactions on Antennas and Propagation*, 36(10), 1443–1454, 1988.

Chapter 2

# Design of monopole circular patch antenna based on DGS for UWB applications

*K. B. Gurumoorthy, A. Arunraja, Cynthia Anbuselvi Thangaraj, R. Chandru, S. Esakki Rajavel, and S. Allwin Devaraj*

## 2.1 INTRODUCTION

Ultra-wideband (UWB) radio technology, which permits short-range, high-bandwidth (BW) communications at extremely low energy levels, can cover a substantial portion of the radio spectrum. Intransigent radar imaging is one of the traditional applications of UWB. The majority of modern applications focus on gathering accurate location and tracking information from sensors. Using a technique known as UWB, data can be transferred over a wide BW (more than 500 MHz). This allows for the transfer of a significant amount of signal energy within the same frequency range without interfering with more traditional NB and carrier wave transmission. The lower threshold of 500 MHz, or 20% of the arithmetic center frequency, is exceeded by the transmitted signal BW for UWB antenna broadcasts.

High-end smartphones started to feature UWB support in the latter part of 2019. Among both academia and industry, UWB technology has captured significant interest, particularly since the FCC officially designated the 3.1–10.5 GHz spectrum for UWB communications. So, it is highly desirable to design such antennas for wireless communications. Short-range wireless communications, UWB technology, microwave radar, remote sensing, and location confirmation applications have seen a significant increase in interest because of the extensive release of the 3.1–10.5 GHz spectrum. The main goals of the UWB antenna design are to improve gain, secure a wide BW, and decrease manufacturing efficiency while maintaining good radiation efficiency. For UWB applications, a monopole-shaped, ring-shaped antenna is suggested. In addition to offering the highest BW in GHz, CPs can be designed with a wide range of lossy properties, show better gain, exhibit desirable electrical and magnetic field strength patterns. In this chapter we propose a simple Monopole Circular Patch (MCP) antenna for UWB applications. UWB resonance characteristics are achieved by placing a circular patch [1,2] on the top side of the substrate and a faulty ground plane with a dumble-shaped slit on the

 DOI: 10.1201/9781003560487-2

bottom surface. With the aid of CST Microwave Studio's 3D modeling and analysis, a planar-modified MCP antenna's design and simulation are proposed [3]. The suggested antenna exhibits nearly omnidirectional radiation patterns, a return loss well below 10 dB, and a VSWR of 2 for the desired frequency range. The disadvantage of the previous technique is a meager antenna gain and BW. It is less capable of handling power. The narrow bandwidth is associated with a tolerance problem. The antenna design is highly complicated and faces insertion loss [4].

## 2.2 LITERATURE SURVEY

Ge et al. [5] propose an antenna with a cavity-back that enables reconfigurable frequency, polarization, and radiation pattern. In addition to switching between two orthogonal Linearly Polarized (LP) and two CP states at the same time, the antenna can also switch its radiation patterns between forward and backward directions [6], as well as tune between three frequency bands for LP and two frequencies for CP. However, it is impossible to avoid insertion losses and the antenna requires a large size and a complex DC biasing circuit.

Wang et al. [7] propose a polarization-agility antenna with high radiation efficiency made of liquid metal (eutectic gallium indium). The suggested layout [8] is based on an aperture-coupled patch antenna made of copper tape and liquid metal alloy. This antenna's polarization is determined by the location of the pressure-driven liquid metal, which is housed in four triangle-shaped cavities [7] made of PET film and Ecoflex, an elastic dielectric material. ECC should be used to oversee the development of liquid metal.

Wang and Chu propose a coaxial dual-tube water antenna with high efficiency and wideband performance [9]. To ensure high radiation efficiency, the distilled water column serves as the loading factor rather than the radiation factor. Fabricated and contrasted are single-tube and dual-tube antennas. The results from simulation and measurement are in surprisingly good agreement [9,10]. The mirror effect and the monopole's electromagnetic radiation may be harmed by the Teflon base's excessive height. In this chapter, we propose a unique [11] single-port quad-polarization-agile aperture-coupled antenna with reversible impedance and a phase converter. Positive-intrinsic-negative (PIN) diodes can be used to manage the reconfigurable converter's impedance and phase changes. Using this converter, a quad-mode switchable feeding network can be created Wideband antennas can be used to improve the polarization-agile antenna's performance [10].

Zhang et al.'s [12] proposed method is further refined using three cascaded slots and four shorting pins on a rectangular patch antenna to perturb the TM21 and TM03 modes and modify their current distributions. Inside this communication, wideband differentially fed patch antennas with twin high-order modes are recommended for dependable high gain less power can be handled by it [13–15]. A symmetrical meandering form resonator is loaded with open-ended identical stubs bent toward each other to minimize circuit size and provide transmission zeros at the target frequencies [16].

Kadam et al. [17] show that combining a CPW feed network, a non-parasitic grounded reflector, and a non-planar GCPW array monopole antenna produces an advantageous beam pattern. Additionally, a miniaturized wide-band bandpass filter operating in the 3–8 GHz frequency band is constructed using Semi-Complementary Split Ring Resonator and Defective Ground Structures (DGS).

Liu et al. [13], by infusing LM into +45 and −45 channels with respect to the x-axis or leaving the channels empty, [18] created Left-Hand Circular Polarization (LHCP), Right-Hand Circular Polarization (RHCP), and Linear Polarization (LP). Liu et al. [13], the concept described by this author is a pixilated antenna construction with a grid of tiny metallic patches and RF switches for reconfigurability [3]. The suggested design prototype is built on a FR4 substrate with a relative dielectric constant of 4.4 and a thickness of 1.6 mm [13]. This antenna can resonate at three distinct frequencies [19].

Hu et al. [20] propose that the antenna supports the X-band and generates dual-band performance. It has been noticed that the return loss is less than 10 dB [16]. The suggested antenna has a gain of 7.37 dB and a gain of 6.37 dB for dual bands. As a consequence, the CSRR antenna's current distribution and radiation plot of the E and H-planes have been provided [20]. The suggested water helical antenna has an overlapping impedance bandwidth (|S11| 10 dB) and an axial ratio (AR) [7] bandwidth ranging from 1.27 to 2.13 GHz, or 50.6% fractional bandwidth, according to Ren [16].

## 2.3 ANTENNA DESIGN

They suggest a straightforward MCP antenna for UWB applications. This can be done by making a Circular Patch Antenna (CPA) on the top sides of the substrate, a flawed ground plane with a dummy slot on the bottom side, and an antenna on the bottom side of the substrate. An MCP antenna's substrate (FR-4 epoxy) measures 42 mm long, 30 mm wide, 0.5 mm high, and 3.4 mm wide, and 0.02 tangent loss (tan). On the substrate's surface, a CPA with a radius of 10 mm is placed, and a 50 microstrip feed line is shown in Figure 2.1. DGS allows for the design of compact structures that have improved bandwidth and gain. Compared to other bandwidth-enhancing methods, manufacturing is also easier. Slots are typically part of the geometry of DGS. Using the proposed MCP antenna, more than two DGS slots are taken into account. UWB with increased antenna performance.

Figure 2.1 depicts the MCP antenna geometry. The MCP antenna's substrate (FR-4 epoxy) measures 48 m in length, 38 mm in width, and 1.6 mm

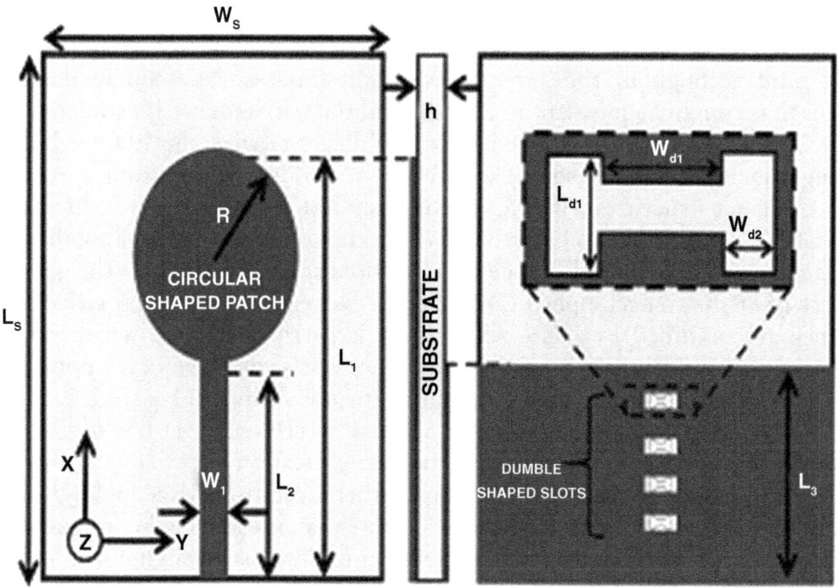


*Figure 2.1* The proposed system is depicted as a block diagram.

in height. It has a *r* value of 4.4 and a tangent loss of 0.22. A CPA with a radius $R$=10mm is put on the substrate's surface and is shown with 50 microstrip feed lines in Figure 2.1. The MCP and strip line lengths are considered as $L1$=40mm, $L2$=20mm with strip line width $W1$=2mm. Defected ground structures (DGSs) on the base of the substrate material are considered to enhance bandwidth. The length of the ground is $L3$=20mm. As shown in Figure 2.1, the DGS plane contains four dumble-shaped slots with dimensions of 1, 1.2, and 0.7mm, respectively.

The effective radius Eq. (2.1) is used to compute the circular path radius, "$R$" (Eq. 2.2)

$$R_{\text{eff}} = \frac{8.79\times 10^{9}}{f_{\text{res}}\sqrt{\varepsilon_r}} \tag{2.1}$$

$$R \frac{R_{\text{eff}}}{\left(1+\frac{2h}{\pi\varepsilon_r R_{\text{eff}}}\left[\ln\left(\frac{1.57R_{\text{eff}}}{h}\right)+1.78\right]\right)^{\frac{1}{2}}} \tag{2.2}$$

Where *r* is the substrate material's dielectric content and free is the substrate material's initial resonant frequency.

### 2.3.1 Defected ground structure (DGS)

Circular and square patches connecting to the feed line are taken into account throughout the antenna evolution process. The square patch's length is roughly equivalent to 20 mm, and the CP diameter is estimated to be 20 mm. According to the reflection coefficient picture, the 50 impedance line-attached CP will resonate with UWB at two operational frequencies.

Compact structures with improved bandwidth and gain are created using DGS. In comparison to other bandwidth and gain enhancement methods, fabrication is also simpler. One or more slots typically make up the geometry of DGS. To accomplish UWB using a better antenna. Microstrip feed lines are modified by DGS below the microstrip feed line, which takes resistance, inductance, and capacitance into account. The corresponding circuit design for a slot with a dumbbell form is shown in Figure 2.2.

Ansys High-Frequency Structure Simulator (HFSS), a 3D EM tool with superior accuracy and speed, is used in this application brief to model antennas. It emphasizes antennas on or near other structures while highlighting various antenna-related applications. Antenna designers who frequently struggle with the challenge of implementing designs throughout an ever-increasing number of frequency bands inside of an ever-shrinking physical space may benefit from HFSS. It achieves this by making use of a variety of simulation techniques and a potent automated adaptive mesh refinement that offers industry-leading accuracy. HFSS simulation is a necessity in the antenna design and integration process given these new technical challenges and the constantly reducing time to market.

### 2.3.2 Design equations for microstrip antenna

A patch's width can be calculated using the following equation:

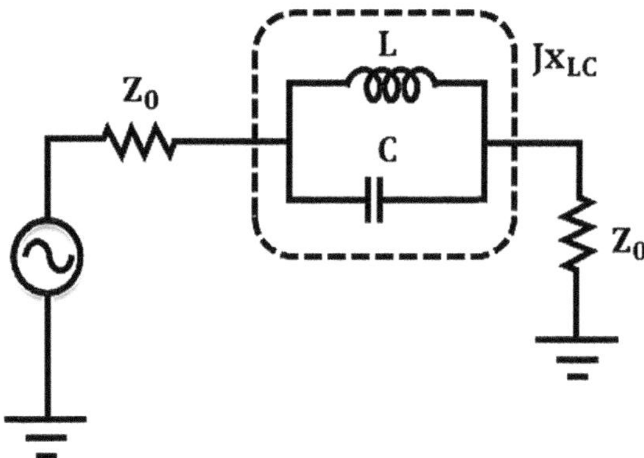


*Figure 2.2* DGS's equivalent circuit schematic with slots shaped like dumbbells.

$$W = \frac{c}{2f_r\sqrt{\frac{\varepsilon_r + 1}{2}}} \tag{2.3}$$

where $W$ is the patch's width, $C$ is light speed, and $r$ is the dielectric substrate value.

The following equation is used to calculate the effective dielectric constant ():

$$\in_{reff} = \frac{\varepsilon_r + 1}{2} + \frac{\varepsilon_r - 1}{2}\left[1 + 12\frac{h}{w}\right]^{-\frac{1}{2}} \tag{2.4}$$

In order to determine the patch's actual length increase ($L$):

$$\Delta L = 0412\text{parent} \tag{2.5}$$

Where $h$=height of the substrate.

In order to determine the patch's length ($L$):

$$L_{\text{eff}} \tag{2.6}$$

$$L = L_{\text{eff}} - 2\Delta L \tag{2.7}$$

## 2.4 RESULT AND DISCUSSION

Using the finite element method and the Ansys HFSSv19 tool, the design and simulation are carried out. The simulated S11 has RL –10 dB and operates with a DGS bandwidth of 8.1 GHz (2.5–10.6 GHz). Dumble slots enhance BW and gain by improvising impedance matching. Numerous wireless applications can benefit from the obtained bandwidth (Figures 2.3 and 2.4). IMT services operate in the 3.5 GHz band; WiMAX applications operate between 3.3 and 3.8 GHz; WLAN applications operate between 5.15 and 5.35 GHz and 5.725 and 5.825 GHz; satellite communications operate between 2–4 GHz and 4–8 GHz, among others.

The applications of various antennas are shown in Figure 2.5, with a focus on antennas on or near other structures. Antenna designers constantly have to execute designs across ever-increasing frequencies across shrinking physical spaces, which can be aided by HFSS. This is accomplished by using a variety of simulation techniques and a potent automated adaptive mesh refinement that offers the highest level of accuracy. HFSS simulation is a necessity in the antenna design and integration process given these new technical challenges and the constantly reducing time to market.

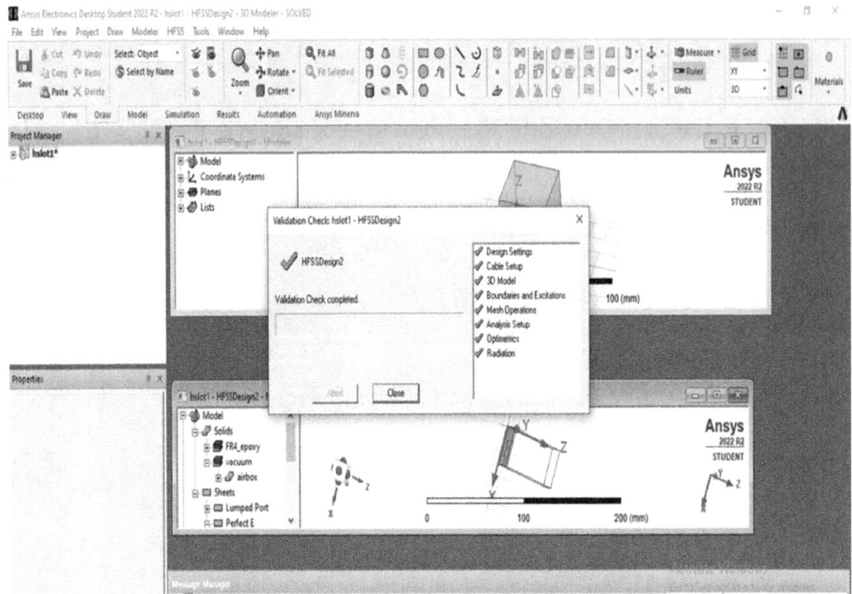

*Figure 2.3* Ansys HFSS v19 view in the simulation of the proposed antenna.

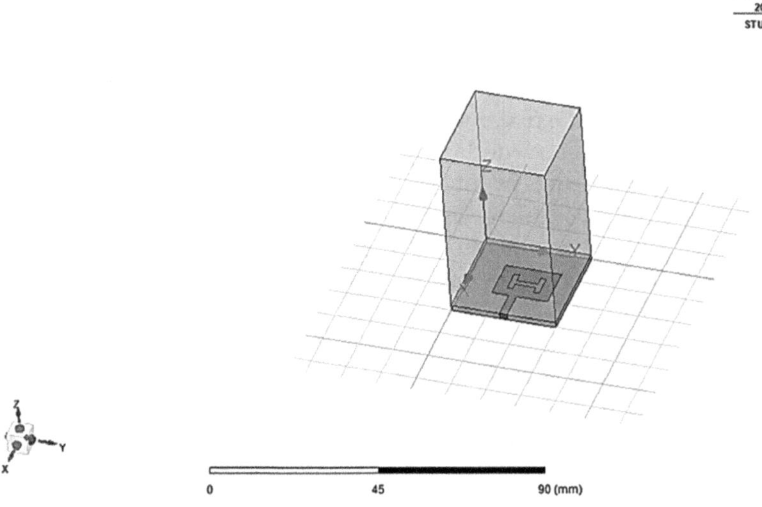


*Figure 2.4* Design of the proposed antenna.

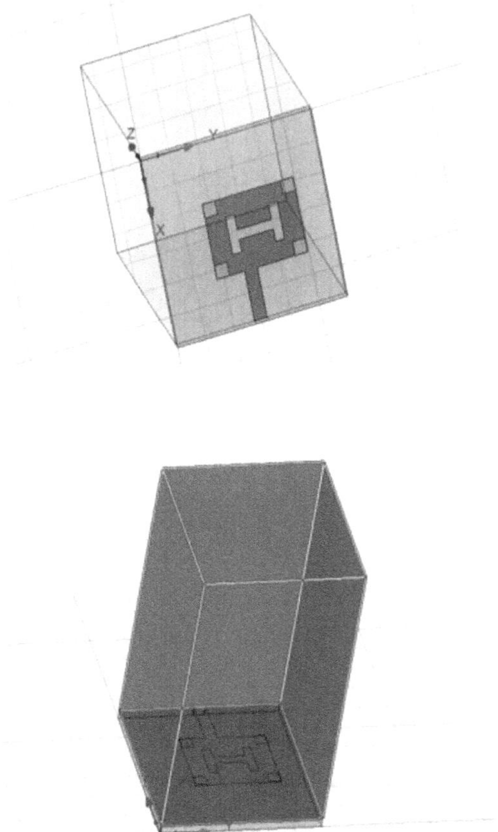

*Figure 2.5* Antenna designs in HFSS software.

One of the methods used most frequently to measure gain is the gain transfer method. Three antennas need to be taken into account with this method.

The equation is used to calculate the proposed antenna's gain. Figure 2.6 depicts the MCP antenna's gain measurement setup. For the 2.9 and 9.1 GHz frequencies are measured peak gains from 8.3 to 7.9 dBi, respectively. The cable and connecting losses are to blame for the discrepancies between the simulated and measured gains.

$$G(\mathrm{AUT}) = G(\mathrm{Ref}) + 10\log 10\left[\frac{P(\mathrm{AUT})}{P(\mathrm{Ref})}\right] \tag{2.8}$$

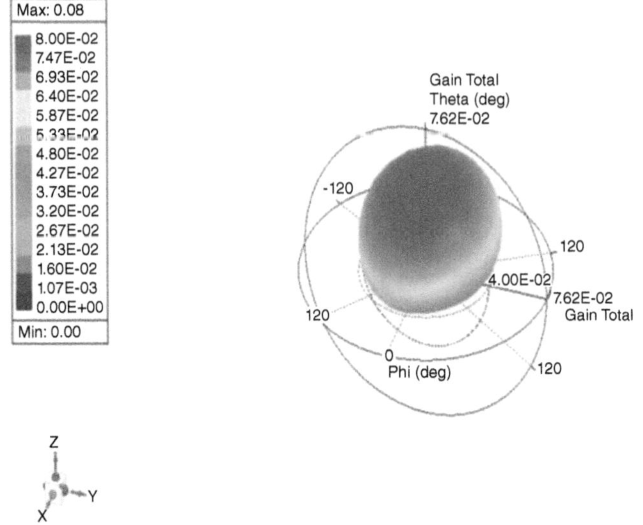


*Figure 2.6* DIR for frequency at 1.20 GHz.

Where,

> *P*(AUT): power received from test antenna.
> *G*(AUT): gain of the antenna being tested.
> *G*(Ref): standard gain from reference antenna.
> *G*(Ref): power received from standard gain antenna.

In Figure 2.7, the *Z* parameter of an antenna is utilized to evaluate its quality factor, which can reveal details about its useful BW. $Z(\text{ant}) = R + jX$, where $R = R(\text{rad}) + R(\text{loss})$, might in some way anticipate the efficiency and losses.

A directional antenna allows for improved performance and less interference from outside sources by radiating or receiving more power in particular directions. When more radiation concentration in a specific direction is desired, directional antennas perform better than dipole antennas—or omnidirectional antennas generally. Figure 2.8 depicts an antenna's return loss, which is the percentage of radio waves that arrive at the antenna input that is rejected in comparison to those that are accepted. For a short circuit, the decibels (dB) are specified to indicate 100% rejection. Consider the antenna in transmit mode.

**A. *S* Parameter Plot**

- This is determined by the incident power to reflected power ratio.
- *S*11/Reflection Coefficient is another name for it.
- It should always be less than –10 dB.
- The suggested antenna has a return loss of –18 dB.

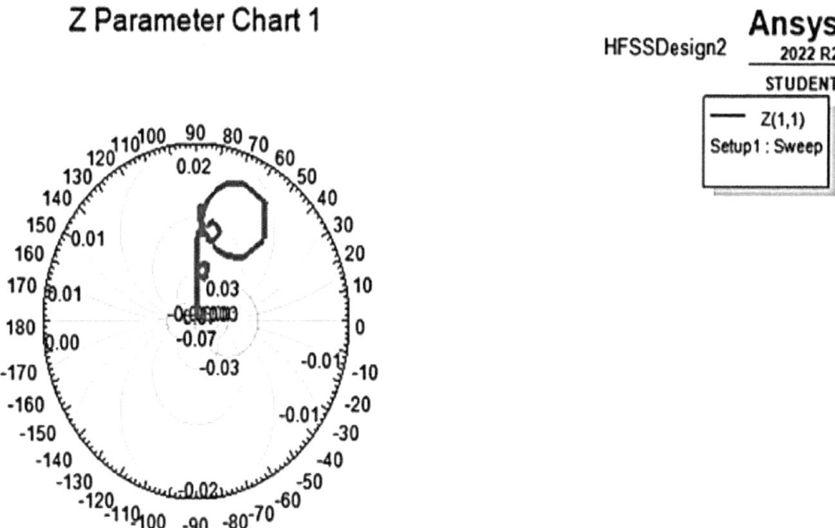


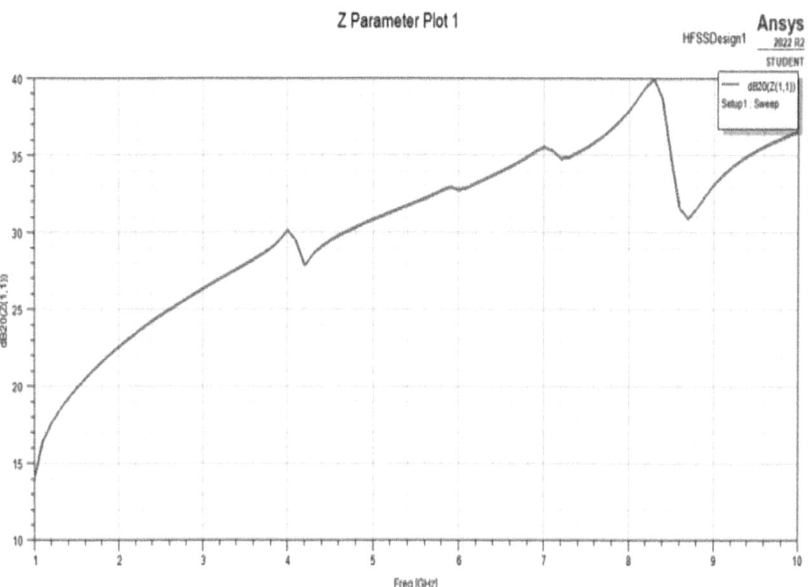


*Figure 2.7* *Z* parameter plot and chart 1.

B. *Y* Parameter Plot

- It is the proportion of incident to reflected power.
- It is also known as the *S*11/Reflection Coefficient.

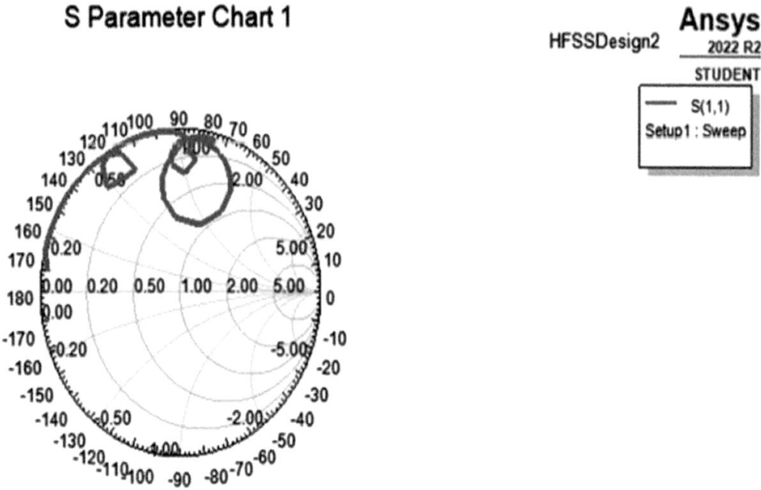


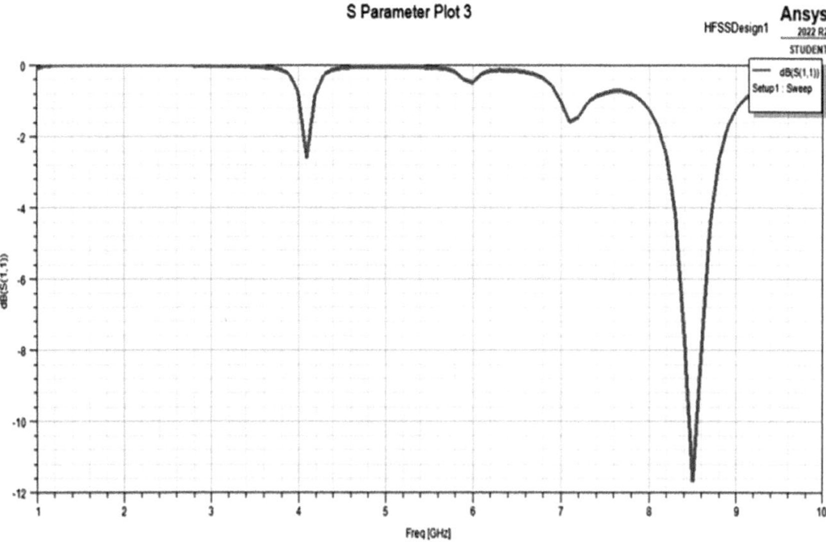


*Figure 2.8* S-Plot and chart 1.

- It should never be less than –10 dB.
- The proposed antenna has a –18 dB return loss.

Figure 2.9 displays the MCP antenna's radiation effectiveness. The observed radiation efficiency is 90.27% at 2.9 GHz and 86.17% at the low frequency of 9.1 GHz. The efficiencies range from 85% to 91%

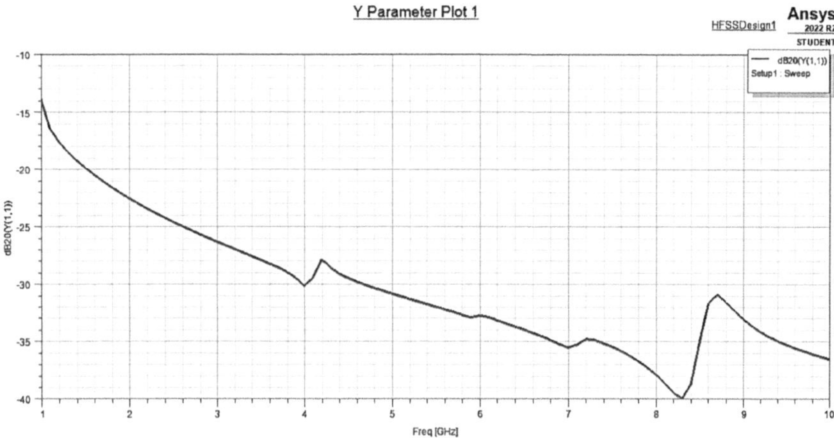


*Figure 2.9* *Y*-parameter plot 1.

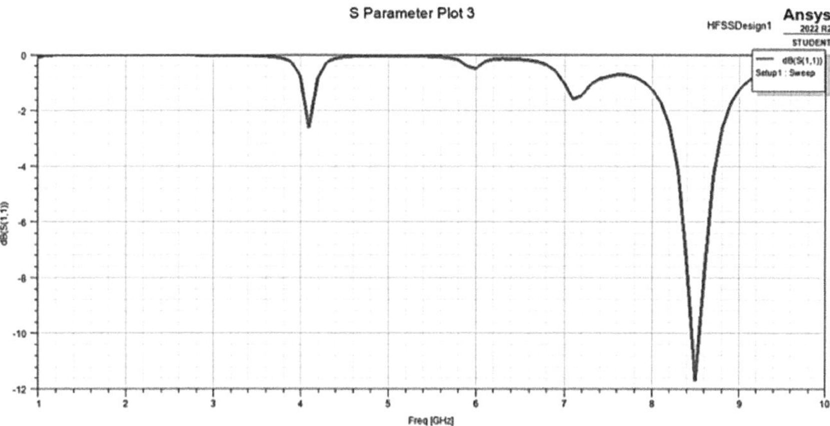


*Figure 2.10* *S*-Parameter plot 1.

for the entire 2.5–10.6 GHz bandwidth. The lossy dielectric substrate materials affect the radiation capability and gain. The proposed MCP antenna gains and radiation efficiencies can be increased by using a lossless microwave substrate material (Figure 2.10).

**C. Gain for Frequency at 8.3 GHz**

It depicts the relative field strength of radio waves radiated by an antenna at various angles. In both the E and H planes, the proposed antenna provides semi-omnidirectional and omnidirectional radiation patterns (Figures 2.11–2.14).

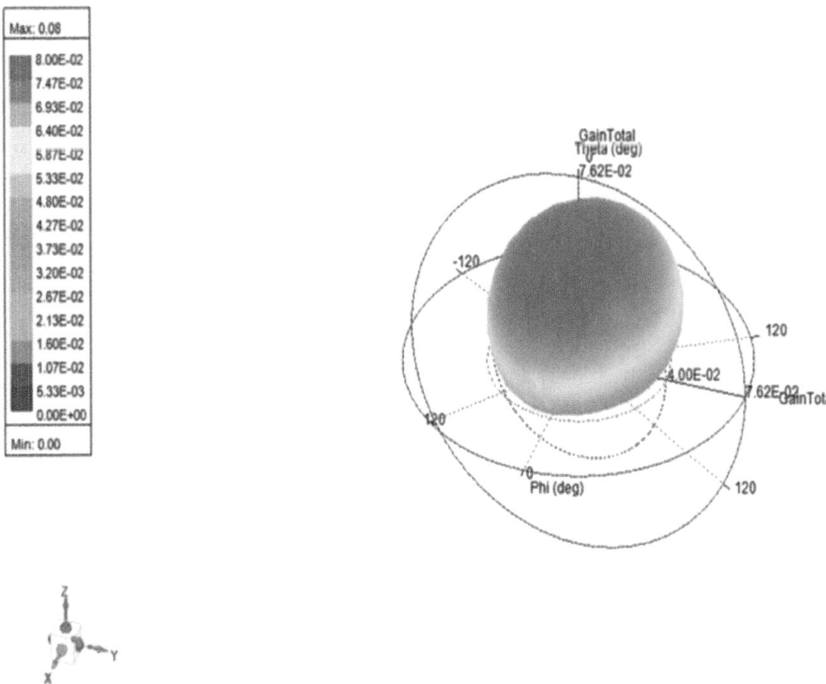


*Figure 2.11* Gain of the proposed antenna.

## 2.5 CONCLUSION

This study proposes a simple MCP antenna for UWB applications. By utilizing the gap section in the circular ring for the reconfigurable network, the recommended approach eliminates the extra space needed for a reconfigurable feed network. In terms of gain characteristics and cross-polarization isolation (15 dB) in the operational band, the antenna outperforms typical reconfigurable antennas. As a result, the antenna is more appropriate for contemporary wireless UWB applications that favor *lp/cp* antenna properties. The proposed design suggests a straightforward MCP antenna for UWB applications. For use in UWB applications, a straightforward $38 \times 48 \times 1.6\,mm^3$ MCP antenna has been successfully integrated into the fr4 substrate. The substrate's bottom has a flawed ground structure, which, increases gains and bandwidth. The ground plane's dumble slot was loaded

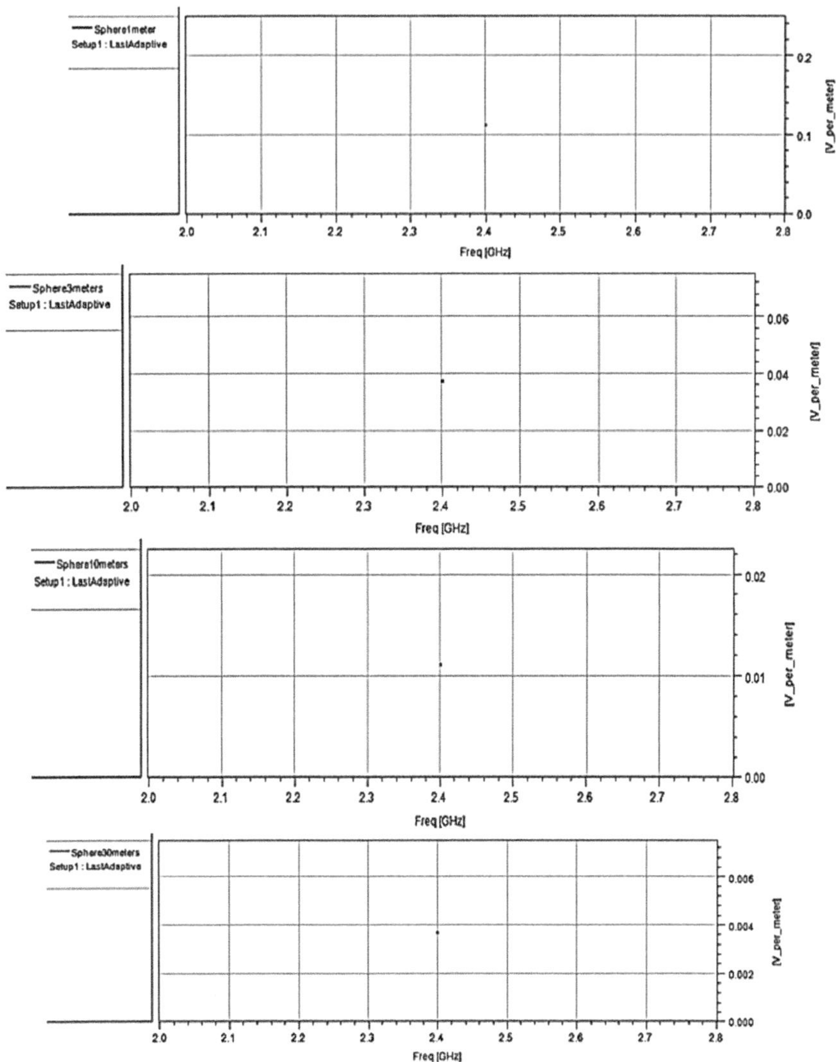


*Figure 2.12* Sphere plots 1, 2, 3, and 4 for freq VS V_per_meter in 1, 3, 10, and 30 m.

correctly, and it successfully accessed the UWB frequency band. With an impedance bandwidth of 8.1 GHz, the MCP antenna offers enhanced gains of 8.4 and 8.2 dBi for wavelengths of 2.9 and 9.1 GHz, respectively.

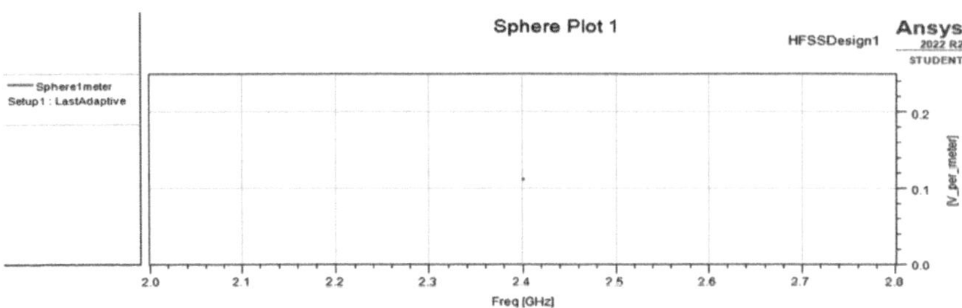


*Figure 2.13* Efficiency.

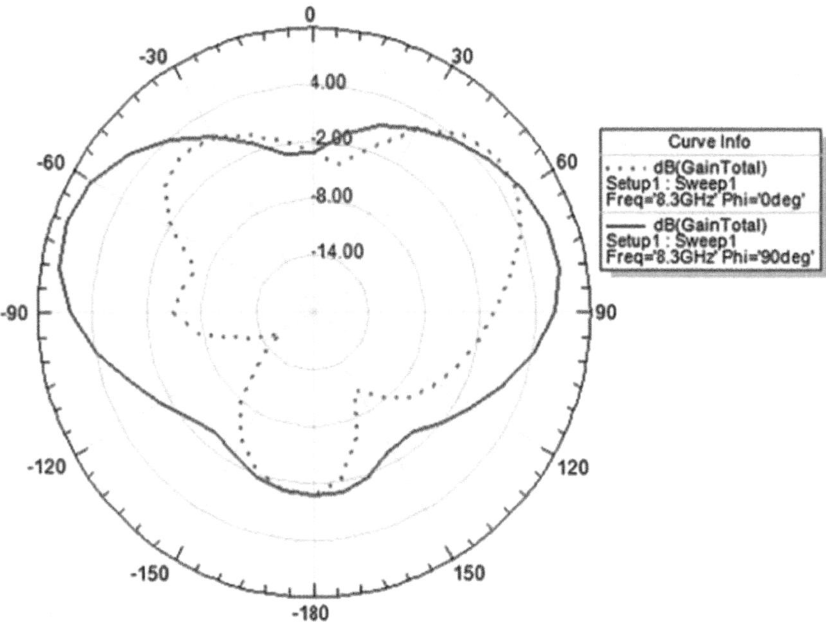


*Figure 2.14* Gain for frequency at 8.3 GHz.

## REFERENCES

1. Cheng, Y. and Dong, Y., 2020. Wideband circularly polarized split patch antenna loaded with suspended rods. *IEEE Antennas and Wireless Propagation Letters*, 20(2), 229–233.
2. Song, Y., 2020. Passive beam-steering gravitational liquid antennas. *IEEE Transactions on Antennas and Propagation*, 68(4), 3207–3212.

3. Huang, Y., 2006. A novel CPS-fed balanced wideband dipole for ultra-wideband applications, *2006 First European Conference on Antennas and Propagation*, France.
4. Hua, C., et al., 2021. Reconfigurable antennas based on pure water. *IEEE Open Journal of Antennas and Propagation*, 2, 623–633. doi: 10.1109/OJAP.2021.3079353.
5. Ge, L., Li, Y. and Wang, J., 2017. A low-profile reconfigurable cavity-backed slot antenna with frequency, polarization, and radiation pattern agility. *IEEE Transactions on Antennas and Propagation*, 65(5), 2182–2189.
6. Chatterjee, J., Mohan, A. and Dixit, V., 2018. Broadband circularly polarized H-shaped patch antenna using reactive impedance surface. *IEEE Antennas and Wireless Propagation Letters*, 17(4), 625–628.
7. Wang, C., Yeo, J.C., Chu, H., Lim, C.T. and Guo, Y.X., 2018. Design of a reconfigurable patch antenna using the movement of liquid metal. *IEEE Antennas and Wireless Propagation Letters*, 17(6), 974–977.
8. Wang, C., Yeo, J.C., Chu, H., Lim, C.T. and Guo, Y., 2018. Design of a reconfigurable patch antenna using movement of liquid metal. *IEEE Antennas and Wireless Propagation Letters*, 68 (11), 7630–7635.
9. Wang, M. and Chu, Q.X., 2018. High-efficiency and wideband coaxial dual-tube hybrid monopole water antenna. *IEEE Antennas and Wireless Propagation Letters*, 17(5), 799–802.
10. Sun, H. and Pan, Z., 2018. Design of a quad-polarization-agile antenna using a switchable impedance converter. *IEEE Antennas and Wireless Propagation Letters*, 18(2), 269–273.
11. Ji, Y., Ge, L., Li, Y. and Wang, J. Wideband polarization agile dielectric resonator antenna with reconfigurable broadside and conical beams, IEEE Conference NCC 2019, Australia.
12. Zhang, X., Hong, K.D., Zhu, L., Bi, X.K. and Yuan, T., 2020. Wideband differentially fed patch antennas under dual high-order modes for stable high gain. *IEEE Transactions on Antennas and Propagation*, 69(1), 508–513.
13. Liu, Y., Wang, Q., Jia, Y. and Zhu, P., 2020. A frequency-and polarization-reconfigurable slot antenna using liquid metal. *IEEE Transactions on Antennas and Propagation*, 68(11), 7630–7635.
14. Zhang, X., Hong, K.-D., Zhu, L., Bi, X.-K. and Yuan, T., 2021. Wideband differentially fed patch antennas under dual high-order modes for stable high gain. *IEEE Transactions on Antennas and Propagation*.
15. Gao, K., Cao, X., Li, T., Yang, H., Li, S.J. and Gao, J., 2022. Wideband single-layer patch antenna with stable high gain using characteristic mode analysis. *IET Microwaves, Antennas & Propagation*, 12, 98–105.
16. Ren, S.-S. Qi, Z. Hu, Z. Shen and W. Wu, 2019. Wideband water helical antenna of circular polarization. *IEEE Transactions on Antennas and Propagation*, 67(11), 6770–6777. doi: 10.1109/TAP.2019.2922846
17. Kadam, A.A., Deshmukh, A.A., Deshmukh, S., Doshi, A. and Ray, K.P., 2019. Slit-loaded circular ultra-wideband antenna for band notch characteristics. In *2019 National Conference on Communications (NCC)*, Bangalore, *2019*. IEEE, pp. 1–6.

18. Abedian, M., Rahim, S., Danesh, S., Khalily, M. and Noghabaei, S., 2013. UWB dielectric resonator antenna with WLAN band rejection at 5.8 GHz. *IEEE Antennas and Wireless Propagation Letters*, 12, 1523–1526.
19. Wong, K.L., Jian, M.F., Chen, C.J. and Chen, J.Z., 2021. Two-port same-polarized patch antenna based on two out-of-phase TM 10 modes for access-point MIMO antenna application. *IEEE Antennas and Wireless Propagation Letters*, 20(4), 572–576.
20. Hu, Z., Wang, S., Shen, Z. and Wu, W., 2017. Broadband polarization-reconfigurable water spiral antenna of low profile. *IEEE Antennas and Wireless Propagation Letters*, 16, 1377–1380.

Chapter 3

# Ultra-wideband circular curve-shaped fractal monopole antenna

*Sheetal Bukkawar and Manjusha Deshmukh*

## 3.1 INTRODUCTION

With the rapid development in Wireless communication technology, the focus is on devices that offer increased functionality, better performance, small size, and low cost. To meet these requirements in mobile radio and wireless communications, microstrip antennas are commonly used. These antennas are low profile, conformable to planar and nonplanar surfaces, simple and inexpensive to fabricate using modern printed-circuit technology, mechanically robust when mounted on rigid surfaces, and compatible with MMIC designs. Besides these, they are very versatile in terms of resonant frequency, polarization, radiation pattern, and impedance [1]. To miniaturize the antenna size with high radiation efficiency, fractal antennas are most suitable. Fractal antennas are the best options to get multiband or wideband with miniaturization.

A fractal is a "rough or fragmented geometric shape" that is generated by starting with a very simple pattern that grows through the application of rules [2]. In many cases the rules to make the figure grow from one stage to the next involve modification, subtraction, or addition to the original figure. The process can be repeated recursively an infinite number of times [3]. Fractal geometry has two important properties: space-filling and self-similarity. The fractal antenna not only has a large effective length, but also features contours of its shape that can generate capacitance or inductance to help match the antenna to the circuit. Applying fractals to the antenna element allows smaller size, multi-band, and broadband properties. The fractal geometry can also improve the radiation pattern of an antenna.

The ultra-wideband (UWB) antenna [4–7] is widely used in many wireless communication systems. The FCC declared unlicensed 3.1–10.6 GHz frequency band as UWB system research. A microstrip UWB antenna was designed with a pentagonal cut radiator patch [8–10]. This antenna yields a better impedance matching 2.5–15 GHz with the conventional coplanar waveguide (CPW) feeding technique. The measured group delay of the

DOI: 10.1201/9781003560487-3

antenna is 0.75 ns. Observing the patterns, the E-plane pattern was asymmetric compared to the H-plane pattern. In Ref. [11], an elliptical radiator patch was used with conventional transmission line excitation. Here, to realize the UWB performance a notched partial ground plane was used. Further, to improve the radiation characteristics and cross-polarization level parasitic elements were used above the partial ground plane. With this configuration, the antenna shows a UWB impedance bandwidth of 3.5–12 GHz and shows a group delay of 1 ns. A pentagon-shaped microstrip radiator was investigated for UWB operation [12], and the antenna structure was modified with Koch fractal. The Koch fractal was also introduced in the ground plane which improved the radiation characteristics. This fractal configuration realized a good impedance bandwidth of 3–12.8 GHz. An octagonal-shaped antenna radiator was investigated for UWB operation [13]. In which a partial ground plane was considered to realize a good impedance bandwidth. The octagonal shape was introduced with step-wise Sierpinski fractal which improves the overall performance of the antenna. The antenna configuration shows a good impedance bandwidth from 2.9 to 12.3 GHz. In Ref [14], a UWB notch band antenna with CPW excitation was presented. Here the ground plane was modified with crown-shaped fractal slot which improved the impedance bandwidth. Here, the band-notched characteristic was achieved with two inverted omega-shaped slots at the radiator patch. Another Band notched UWB antenna was realized in Ref [15]. The antenna radiator was implemented with a rectangular, tree-like fractal structure. The fractal radiator was excited with a microstrip-fed line, and a notched partial ground plane was used to improve the impedance bandwidth.

In another framework, a Cylindrical Dielectric Resonator Antenna structure was incorporated to realize UWB operation [16]. The antenna configuration included a snowflake Koch fractal and later a fractal ring was created on the fractal geometry to improve the performance. However, the desired UWB response was realized but the radiation patterns did not meet the realistic expectations. A coaxial excited circular patch antenna was investigated in Ref [17]. Here, in the beginning, the circular radiator shows a triple band resonance. Later the radiator was modified with fractal shapes at the diagonal corner which realized a good UWB performance. Another circular radiator patch was investigated for UWB performance in Ref [18]. Here the antenna radiator was modified with a tree-like fractal structure where a partial ground plane was used to promising UWB performance. However, the radiation patterns were not symmetric in UWB bands. In Ref [19], a circular radiator patch was studied with different slotted techniques such as triangular, square, annular, and Koch fractal shapes. The modified square slot with Koch fractal gave the UWB performance. Even though a good radiation pattern was observed throughout the band, asymmetricity was observed at the edge frequencies. A CPW-fed fractal slotted radiator patch was investigated in Ref [20] for UWB performance. Here a superstrate layer was incorporated to improve the gain response, however the patterns in E-plane were asymmetric in nature.

A cactus-shaped UWB mono pole antenna was studied to realize better UWB performance [21]. This antenna was designed on a special type of low-loss, ultra-thin material called liquid crystal polymer. This antenna promises a good UWB bandwidth with a symmetric radiation pattern at all frequencies. A multinotched UWB circular radiator patch was investigated for UWB performance [22]. This configuration used a CPW feeding mechanism which was designed to give omni-directional patterns. In Ref [23], a circular arc radiator patch with CPW-fed was investigated for UWB performance. This antenna had gone through a few iterative steps, in which the radiator patch was introduced with elliptical slots. The final configuration showed a good UWB performance with a low delay response. Another circular radiator patch with microstrip line-fed was examined for UWB performance [24]. Here the radiator patch was introduced with Inscribed Fibonacci Circle Fractal which drastically reduced the overall antenna aperture size with promising UWB performance. Along with this, the structure used a partial ground plane with a modified notched configuration which improved the UWB performance.

In Ref [25], a microstrip-fed fractal slotted antenna was investigated for UWB and multipurpose IoT applications. Here, a trident shape fractal was introduced in the radiator patch and later the antenna patch was configured as an EBG structure. As a convenience, the antenna used a partial ground plane which commits a good UWB performance with impedance bandwidth from 1.59 to 13.31 GHz. However, the antenna gave a moderate gain of 2.52 dBi. A modified dipole antenna was investigated for UWB operation [26]. In this antenna the radiator patch was introduced with a hexagonal Sierpinski grid fractal gasket which improved the overall performance of the antenna. This antenna realized a very small UWB bandwidth of 1.25–4.95 GHz. In Ref [27], a compact microstrip antenna radiator was discussed, in which CPW excitation was used. In this antenna, the ground plane was modified to a semi-tapered shape which commits a good impedance matching of 3–11.2 GHz. This antenna shape was introduced with self-similar fractal-like corrugated fractal which improves the antenna performance. Another CPW-fed microstrip antenna was investigated for UWB operation [28]. Here, a circular radiator patch was examined with a fractal configuration. This antenna structure was eventually introduced with circular fractal shapes which improved the UWB performance. This antenna showed a good UWB impedance matching of 3.1–10 GHz.

In this chapter, a circular curve-shaped monopole fractal antenna is proposed. The antenna offers 2.65–12.1 GHz impedance bandwidth. The radiation patterns are omnidirectional in lower UWB and directional in upper UWB. The following sections deal with the antenna geometry, simulation, and experimental results [29, 30].

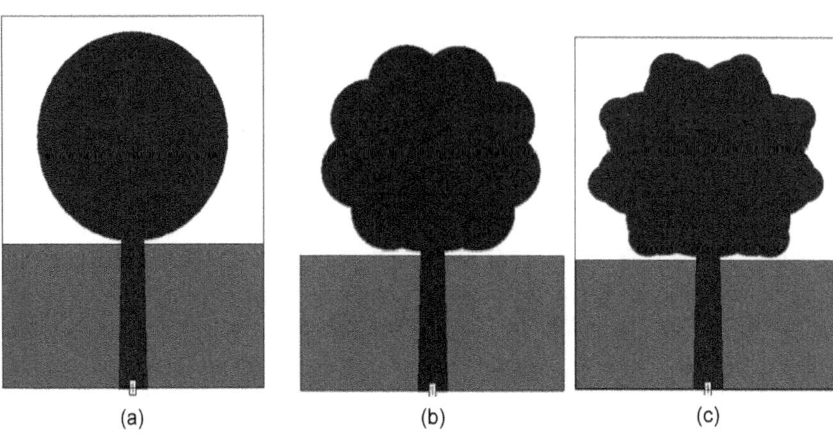


*Figure 3.1* Different stages of antenna configuration. (a) Primary antenna/initiator (antenna 1), (b) iteration 1 (antenna 2), and (c) iteration 2/final antenna (antenna 3).

## 3.2 ANTENNA GEOMETRY AND DESIGN THEORY

Initially, a circular microstrip patch is considered as a primary radiator of the antenna. In which a regular transmission line is placed at the bottom side of the radiator patch for the purpose of excitation. As discussed, in this primary antenna configuration also called an initiator, a partial ground plane is used to acquire UWB operations. This antenna configuration also does not include a slit at the ground plane. The primary antenna configuration of circular curve-shaped UWB antenna is shown in Figure 3.1a.

The discussed primary antenna structure is implemented on a substrate of length $L$ and width of $W$, on which the circular radiator patch is drawn with radius $R$. This structure uses the partial ground of length $L_g$ and width $W_g$. The initial dimension of the circular radiator is defined from the predefined mathematical equation presented in Eq. (3.1),

$$f_L = \frac{7.2}{2.25R + g}\text{GHz} \tag{3.1}$$

Later to improve the bandwidth response of the proposed UWB antenna a few iterations have been carried out. In iteration 1, another circular shape is introduced to form the fractal shapes at the edges of the circular radiator as shown in Figure 3.1b. In the second iteration, again circular arcs are introduced in the existing geometry to further improve the performance as shown in Figure 3.1c which looks like a flower shape. The design parameters of the proposed circular curve-shaped antenna are illustrated in Figure 3.2.

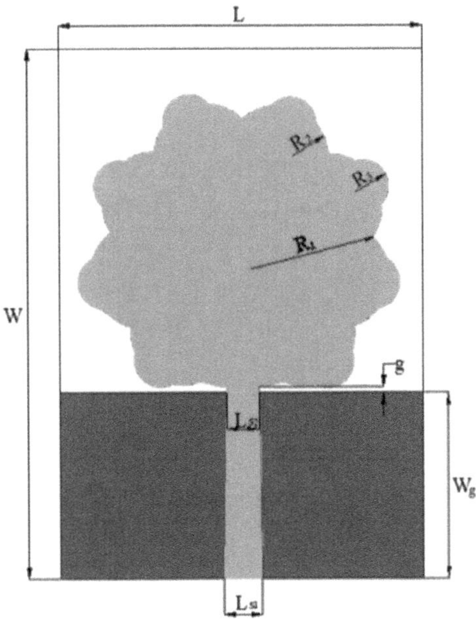


*Figure 3.2* Final geometry of the proposed UWB antenna.

*Table 3.1* Optimized dimensions of the circular curve-shaped UWB antenna

| *Parameter* | *Dimensions (mm)* |
|---|---|
| *L* | 28 |
| *W* | 39 |
| *Lg=L* | 28 |
| *Wg* | 14 |
| *G* | 0.3 |
| *R1* | 10.2 |
| *R2* | 4 |
| *R3* | 2 |
| *Ls1* | 3 |
| *Ls2* | 2.4 |

The optimized dimensions of the proposed circular curve shape UWB antenna are mentioned in Table 3.1.

The final return loss characteristics of the proposed circular curve shape UWB antennas are presented in Figure 3.3 along with the band

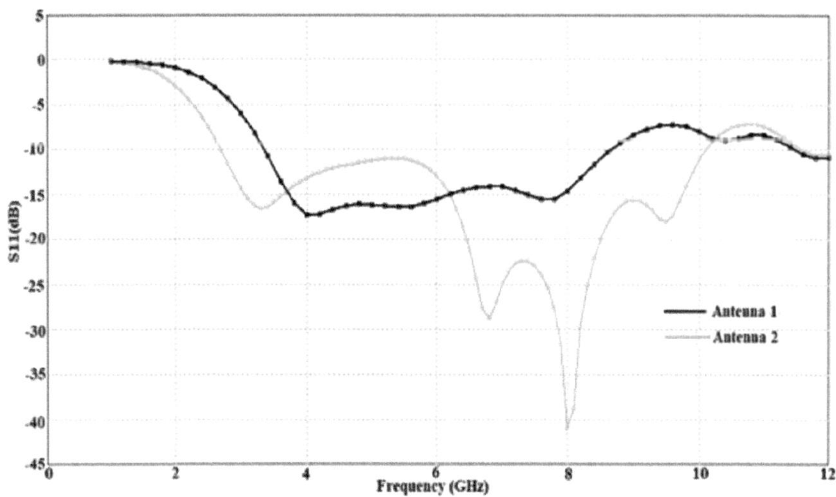


*Figure 3.3* S11 parameter of the circular curve shape UWB antennas with different configurations.

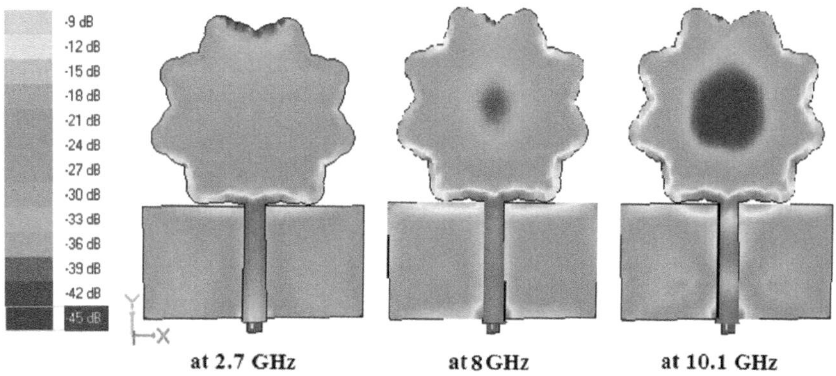


*Figure 3.4* Surface current distribution of designed circular curve shape antenna at various frequencies.

identification. The optimized fractal circular curve shape antenna yields a good impedance bandwidth of 2.65–10.1 GHz.

The surface current distribution shown in Figure 3.4 indicates that as the frequency increases the current density is more at the edges of the antenna and much less at the center of the patch.

## 3.3 EXPERIMENTAL VALIDATIONS OF DESIGNED ANTENNA

The final structure of the circular curve-shaped fractal UWB is fabricated on an FR4 substrate with copper cladding on both sides. The fabrication follows the photolithography process where the antenna geometry is converted into negative film and exposed to UV rays. After that chemical etching was done to get the proper antenna geometry. The fabricated antenna prototype is shown in Figure 3.5, which displays the antenna radiator geometry and the partial ground plane.

Figure 3.5 shows the radiator patch and the partial ground plane of the fabricated circular curve-shaped fractal antenna. The fabricated antenna is tested using Vector Network Analyzer (VNA). In the very beginning stage, VNA is calibrated using short/open/through methods for the frequency range. After which the return loss of the final antenna is measured. The measured return loss is compared with the simulation result as illustrated in Figure 3.6.

The measured return loss shows a good agreement with the simulated result acquired from the EM solver.

In the subsequent stage, the radiation performance of the proposed circular curve shape fractal UWB antenna is observed. Typically, the far-field radiation patterns are observed and presented below with a suitable explanation. The radiation pattern at different planes with respect to corresponding theta and phi is observed. The radiation patterns are observed

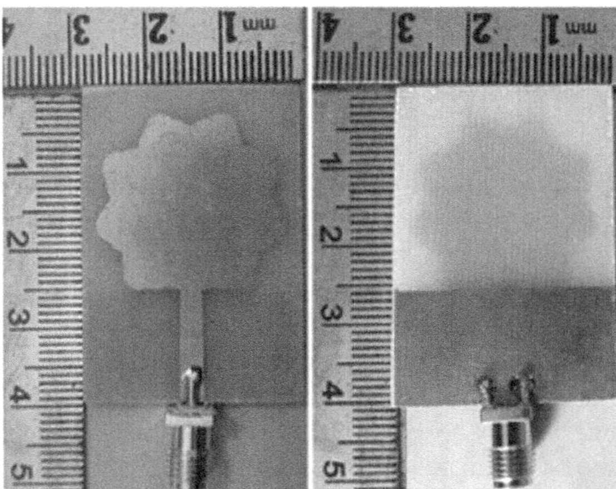

*Figure 3.5* Radiator patch and Ground plane.

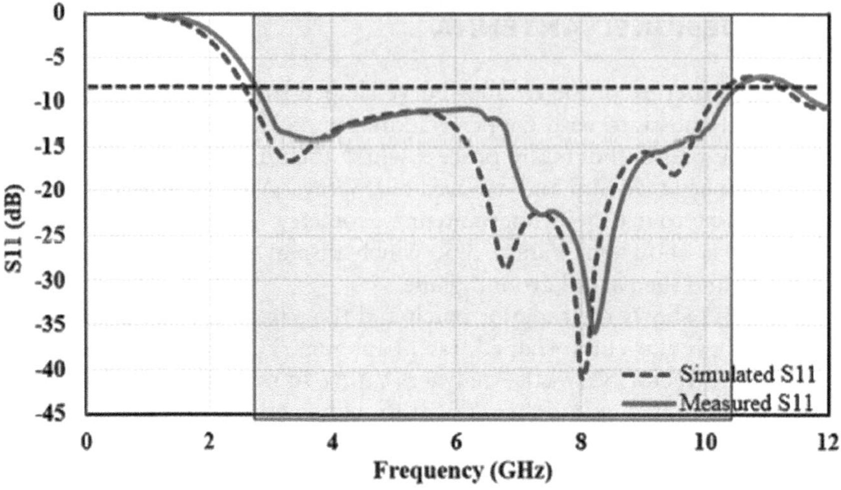


*Figure 3.6* Measured and simulated return loss of the flower-shaped antenna.

in both the E-Plane and H-Plane at specific frequency points, as demonstrated in Figure 3.7. In addition to this the cross-polarization behavior of the antenna with respect to co-polarization patterns is observed and plotted in Figure 3.7.

In Figure 3.7, the E-plane radiation patterns are plotted on the left side and H-plane patterns are plotted on the right side. Co-polarization patterns are marked in red color, while the cross-polarization patterns are marked with a blue line. It has been observed that the co-polarization patterns are close to 0 dB, and cross-polarization patterns are well around –40 dB which confirms good radiation characteristics at different frequency points for the UWB circular curve-shaped antenna. The radiation pattern over the UWB bandwidth is symmetric, as shown in Figure 3.7. However, it loses its symmetry at high frequency i.e. the beam width of the pattern is narrowed, and some beam squint is observed.

Furthermore, the directivity and gain response of the designed circular curve shape UWB antenna is realized during the EM simulation. It is observed that the directivity of the UWB antenna linearly increases with the iteration and the final antenna configuration delivers a good directivity as shown in Figure 3.8. Figure 3.8a shows the directivity of antenna 1 where a sudden peak is observed at a higher frequency which may affect the UWB response. Figure 3.8b shows the directivity of antenna 2/final antenna which has a uniform response throughout the band. Although a small deviation of 0.5 dB is observed, it can be ignored.

The gain response of the designed antenna with different iterations has been observed and presented in Figure 3.9. In Figure 3.9a, the gain response of antenna 1 is presented and in the gain response of antenna 2 (final antenna) is presented. It is observed the gain response is improved by around 0.5 dB at higher frequency.

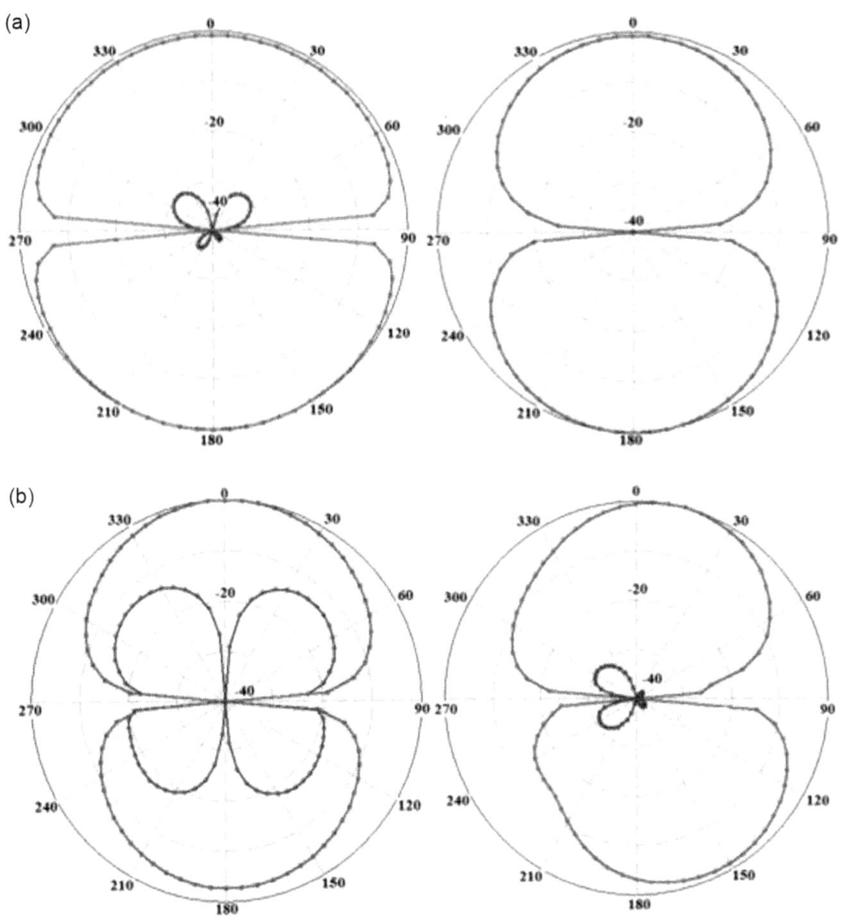


*Figure 3.7* Radiation patterns of flower shape UWB antenna. (a) Radiation pattern at 2.7 GHz, (b) Radiation pattern at 5.2 GHz, (c) Radiation pattern at 6 GHz, and (d) Radiation pattern at 7 GHz.

*(Continued)*

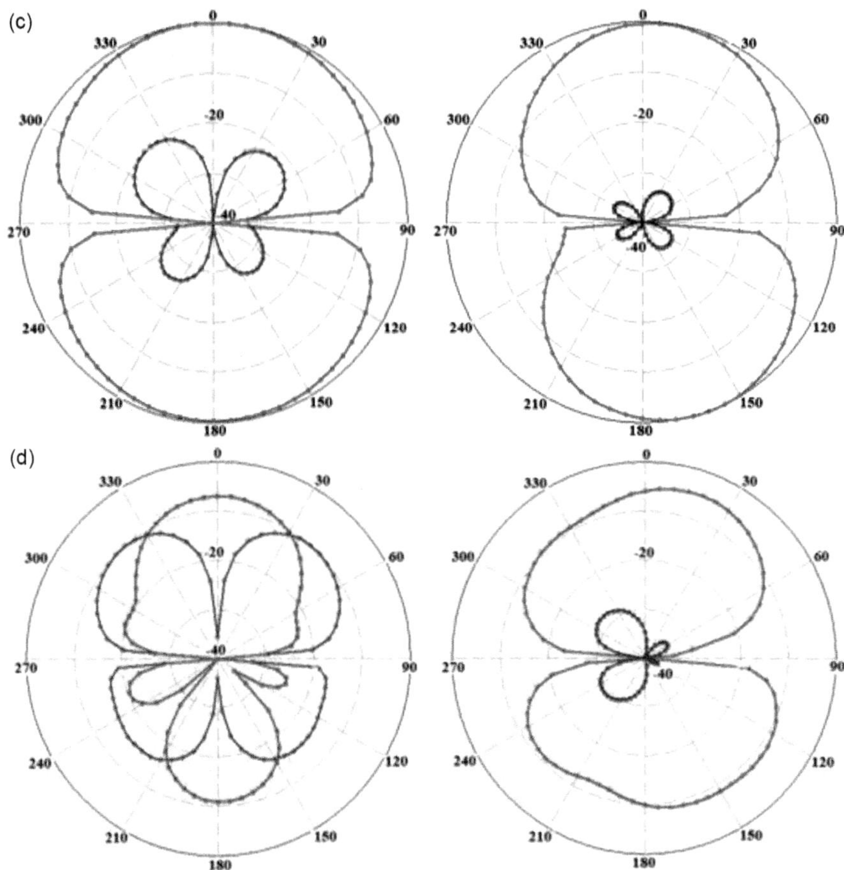


*Figure 3.7 (Continued)* Radiation patterns of flower shape UWB antenna. (a) Radiation pattern at 2.7 GHz, (b) Radiation pattern at 5.2 GHz, (c) Radiation pattern at 6 GHz, and (d) Radiation pattern at 7 GHz.

## 3.4 CONCLUSION

The demonstrated circular curve shape UWB antenna shows a good bandwidth of 2.65–10.1 GHz. The simulation results indicate that the impedance bandwidth i.e. $S11 > -10\,dB$ lies within this frequency band. The proposed antenna is constructed and tested. The measured result follows the simulation results. Also, the antenna shows a symmetric radiation pattern throughout the bandwidth at different frequency points. The gain response of this antenna is better compared to the other UWB antenna configurations. However, the size of the antenna goes up due to the introduction of circular shapes on the edges. Also, due to the introduction of the circular shapes, the aperture size increases which tends to raise the resistive loss which reduces the antenna efficiency.

(a)

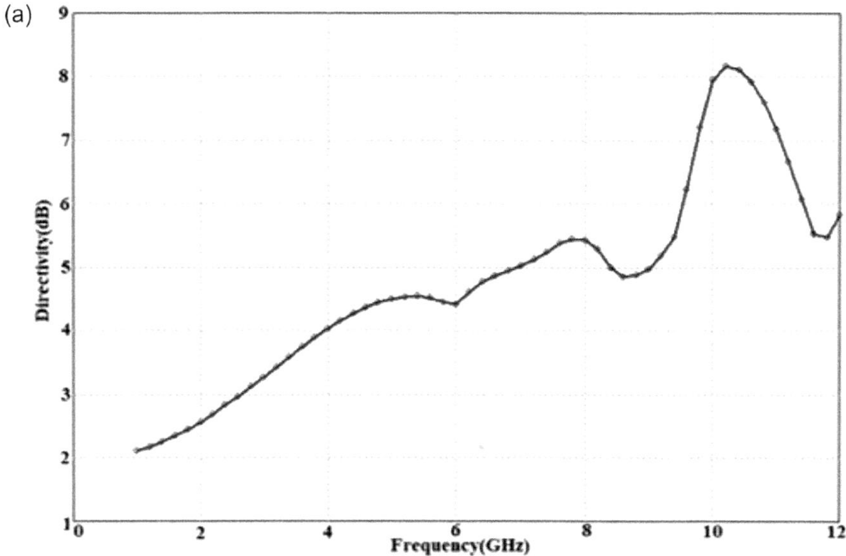


(b)

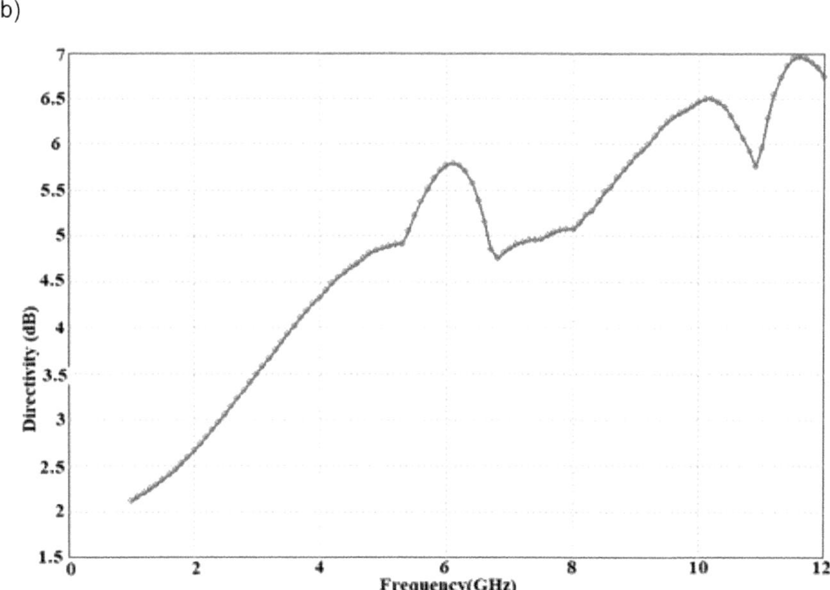


*Figure 3.8* Directivity responses of the antenna with different configurations. (a) Antenna 1 and (b) Antenna 2.

(a)

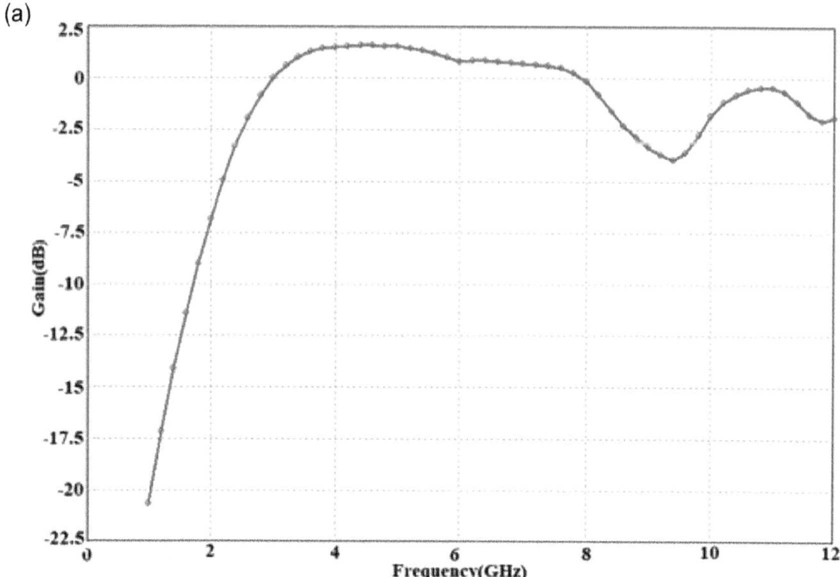


(b)

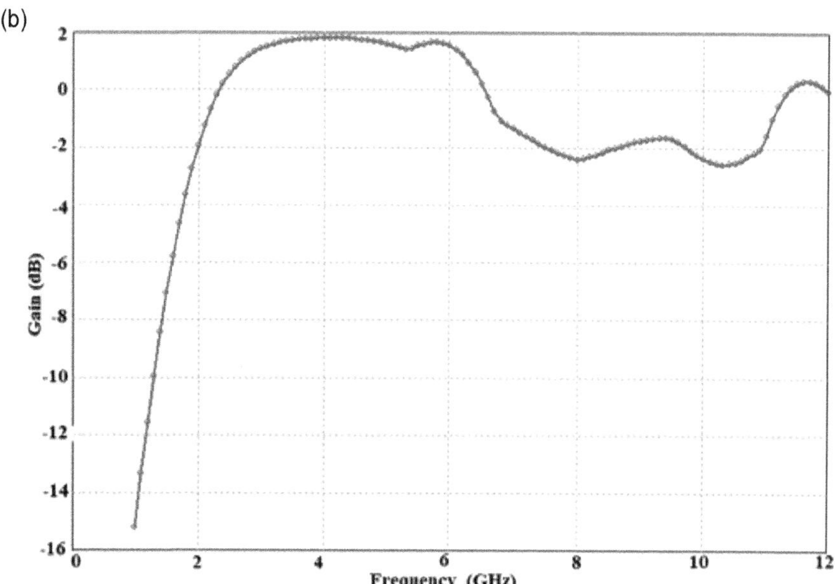


*Figure 3.9* Gain responses of the flower shape UWB antenna. (a) Antenna 1 and (b) Antenna 2.

## REFERENCES


1. I. F. Akyildiz, W. Su, Y. S. Subramaniam, and E. Carirci, "Wireless sensor networks: A survey", *Elsevier Journal on Computer Networks*, Vol. 38, No. 4, pp. 393–442, 2002.
2. Base camp, White Paper, *Internet of Things: Wireless Sensor Networks*, International Electrotechnical Commission, Geneva, Switzerland, 2014.
3. Y. T. Lo, D. Solomon, and W. F. Richards, "Theory and experiment on microstrip antennas", *IEEE Transaction on Antennas and Propagation*, Vol. AP-27, No. 2, pp. 137–145, 1979.
4. D. Guha and Y. M. N. Antar, *Microstrip and Printed Antennas New Trends Techniques & Application*, John Wiley & Sons, Inc Publication, UK, second edition, Chapter 2, 2011.
5. W. L. Stutzmanet and G. A. Thiele, *Antenna Theory and Design*, John Wiley & Sons, Inc Publication, USA, third edition, 2013.
6. G. Kumar and K. P. Ray, *Broadband Microstrip Antennas*, Artech House, Boston, MA; London, 2003.
7. Z. Tang, X. Wu, J. Zhan, Z. Xi, and S. Hu, "A novel miniaturized antenna with multiple band-notched characteristics for UWB communication applications", *Journal of Electromagnetic Waves and Applications*, Vol. 32, No. 15, pp. 1961–1972, June 2018.
8. A. A. Bhoot, S. A. Memon, A. Ahmed, and S. Hussain, "Comparative study of microstrip patch antenna with different shapes and its application," *2019 2nd International Conference on Computing, Mathematics and Engineering Technologies (iCoMET)*, Sukkur, Pakistan, 2019, pp. 1–8.
9. C. Sim, W. Chung, and C. Lee, "Compact slot antenna for UWB applications", *IEEE Antennas and Wireless Propagation Letters*, Vol. 9, pp. 63–66, 2010.
10. H. Oraizi and S. Hedayati, "Miniaturized UWB monopole microstrip antenna design by the combination of Giusepe Peano and Sierpinski Carpet fractals", *IEEE Antennas and Wireless Propagation Letters*, Vol. 10, pp. 67–70, 2011.
11. J. M. Patel, S. K. Patel, and F. N. Thakkar, "Design of S-shaped multiband microstrip patch antenna", *IEEE Proceedings*, Vol. 2158, No. 24, 2017.
12. J. Kim, T. Jung, H. Ryu, J. Woo, C. Eun, and D. Lee, "Compact multiband microstrip antenna using inverted-L- and T-shaped parasitic elements", *IEEE Antennas and Wireless Propagation Letters*, Vol. 12, pp. 1299–1302, 2013.
13. S. C. Basaran, U. Ulgun, and K. Sertel, "Multiband monopole antenna with complementary split-ring resonators for WLAN and WiMAX applications", *IET Electronics Letters*, Vol. 49, No. 10, pp. 636–638, 2013.
14. K. Lin, C. Lin, and Y. Lin, "Simple printed multiband antenna with novel parasitic-element design for multistandard mobile phone applications", *IEEE Transactions on Antennas and Propagations*, Vol. 61, No. 1, pp. 488–491, 2013.
15. C. W. Jung, I. Kim, Y. Kim, and Y. E. Kim, "Multiband and multifeed antenna for concurrent operation mode", *IET Electronics Letters*, Vol. 43, No. 11, 2007.
16. H. Elsadek, and D. M. Nashaat, "Multiband and UWB V-shaped antenna configuration for wireless communications applications", *IEEE Antennas and Wireless Propagation Letters*, Vol. 7, pp. 89–91, 2008.

17. B. Manimegalai, S. Raju, and V. Abhaikumar, "A multifractal cantor antenna for multiband wireless applications", *IEEE Antennas and Wireless Propagation Letters*, Vol. 8, pp. 359–362, 2009.
18. J. Liu and W. Yin, "A compact inter-digital capacitor-inserted multiband antenna for wireless communication applications", *IEEE Antennas and Wireless Propagation Letters*, Vol. 9, pp. 922 925, 2010.
19. T. Zhang, R. Li, G. Jin, G. Wei, and M. M. Tentzeris, "A novel multiband planar antenna for GSM/UMTS/LTE/Zigbee/RFID mobile devices", *IEEE Transactions on Antennas and Propagations*, Vol. 59, No. 11, pp. 4209–4214, 2011.
20. H. F. Abutarboush, R. Nilavalan, S. W. Cheung, and K. M. Nasr, "Compact printed multiband antenna with independent setting suitable for fixed and reconfigurable wireless communication system", *IEEE Transactions on Antennas and Propagations*, Vol. 60, No. 8, pp. 3867–3874, 2012.
21. J. Dong, X. Yu, and L. Deng, "A decoupled multiband dual-antenna system for WLAN/LTE smartphone applications", *IEEE Antennas and Wireless Propagation Letters*, Vol. 16, pp. 1528–1532, 2017.
22. H. F. Abutarboush, H. Nasif, R. Nilavalan, and S. W. Cheung, "Multiband and wideband monopole antenna for GSM 900 and other wireless applications", *IEEE Antennas and Wireless Propagation Letters*, Vol. 11, pp. 539–542, 2012.
23. J. Anguera, A. Andujar, and C. Garcia, "Multiband and small coplanar antenna system for wireless handheld devices", *IEEE Transactions on Antennas and Propagations*, Vol. 61, No. 7, pp. 3782–3789, 2013.
24. A. Mehdipour, I. D. Rosca, A. Sebak, C. W. Trueman, and S. V. Hoa, "Full-composite fractal antenna using carbon nanotubes for multiband wireless applications", *IEEE Antennas and Wireless Propagation Letters*, Vol. 9, pp. 891–894, 2010.
25. Z. Wang, L. Z. Lee, D. Psychoudakis, and J. L. Volkais, "Embroidered multiband body: Worn antenna for GSM/PCS/WLAN communications", *IEEE Transactions on Antennas and Propagations*, Vol. 62, No. 6, pp. 3321–3329, 2014.
26. V. Rajeshkumar and S. Raghavan, "A compact matematerial inspired triple band antenna for reconfigurable WLAN/WiMAX applications", *International Journal of Electronics and Communication*, Vol. 69, pp. 274–280, 2015.
27. S. Verma and P. Kumar, "Compact arc-shaped antenna with binomial curved conductor-backed plane for multiple wireless applications", *IET Microwaves, Antennas & Propagation*, Vol. 9, No. 4, pp. 351–359, 2015.
28. Y. F. Cao, S. W. Cheung, and T. I. Yuk, "A multiband slot antenna for GPS/WiMAX/WLAN system", *IEEE Transactions on Antennas and Propagations*, Vol. 63, No. 3, pp. 952–958, 2015.
29. A. Kunwar, A. K. Gautam, and K. Rambabu, "Design of a compact U-shaped slot triple band antenna for WLAN/WiMAx applications", *International Journal of Electronics and Communication*, Vol. 71, pp. 82–88, 2017.
30. R. Kumar and P. N. Chaubey, "On the design of inscribed pentagonal-cut fractal antenna for ultra wideband applications", *Microwave and Optical Technology Letters*, Vol. 53, pp. 2828–2830, 2011.

Chapter 4

# Design and analysis of a UWB antenna with reduced SAR for early detection of brain tumour

*B. A. Sapna, K. Ramasamy, M. Gopi Krishnan, S. Dhanyasree, and S. Jegadesh*

## 4.1 INTRODUCTION

Advancement in antenna design is increasing for healthcare applications from the COVID-19 arrival due to vulnerability to virus attacks by patients and doctors. Wireless body area networking (WBAN), including on-body communication, off-body communication and in-body communication, has been developed with devices placed on the human body, external to the body or implanted in the body for various medical applications. In WBAN biomedical antennas are either communicating antennas that transfer information from the body to remote servers or locations where doctors can access data or communicate between nodes of the network placed at different parts of the body, or as detecting antennas that detect the presence of any abnormalities in the body. One of the most common disorders of the neurological system is a brain tumour that impairs people's ability to operate normally. A tumour is an abnormality developed as enlargements in cells due to multiplicative cell growth. Brain cancer is a serious deadly disease and if no proper diagnosis is performed will create a significant negative impact on the quality of life, of the patient and their loved ones. There are a variety of diagnosing techniques available in the medical industry including skull X-rays, computer tomography (CT) or magnetic resonance imaging (MRI) but their expense is on the upper end for the common man [1,2]. Hyperthermia [3,4] is one of the more promising therapeutic approaches for malignant brain tumours which utilize localized heating to kill the cancer cells of the brain without damaging normal cells. Imaging systems with neural technic and edge detection are applicable for brain tumour detection [5–8]. Microwave-based imaging techniques are becoming attractive in the medical industry to treat cancer cells [9,10]. In microwave imaging antennas are used to detect tumours based on variation in dielectric properties of the human tissues and malignant tissues.

There are certain challenges in designing antennas for tumour detection. Brain tumours occur anywhere in the brain either on the top layer or deep beneath. Thus, choice of antenna design is also made based on frequency, penetration depth, biocompatibility and sensitivity [11]. Low-frequency

DOI: 10.1201/9781003560487-4

antennas can penetrate deep inside the tissues for localization of the tumour and should be designed on flexible biocompatible materials to improve the efficiency of the imaging system. When resolution is a major factor array antennas or beam-forming antennas are employed for tumour detection using microwave imaging. There are a variety of antennas used for tumour detection from narrow-band to ultra-wideband antennas with different frequencies of operation. An array of antennas with EBG and metamaterials are also utilized to cover the entire head to precisely locate the tumour [12–14] which also reduces the specific absorption rate(SAR). UWB antennas have good penetration depth with wider coverage and higher gain [15–17]. This chapter illustrates the design, simulation and working of a UWB antenna and how it is used for detecting brain tumours by simulation with a rectangular head equivalent model and available head phantom HFSS model. The reflection response of the antenna or the current distribution can be used to identify the presence of malignant tissues. Investigation of SAR is also described here as it is one of the important parameters for radiating elements affecting human tissue.

## 4.2 DESIGN METHODOLOGY – ANTENNA GEOMETRY EVOLUTION

The proposed UWB monopole antenna consists of an annular ring with an outer and inner radius of 9.5 and 5 mm printed on top of a 2 mm thick silicon rubber substrate. There are two square slots placed diagonally on the lower and upper half of the circular ring. A partial ground is placed below the substrate. The ground plane also consists of two square slots on either side of the feed line and a step cut at the centre. Copper is used to make the ground plane and patch with a thickness of 0.05 mm. The dielectric material used is silicon rubber which has a relative permittivity of 2.89 and loss tangent 0.023. The overall size of the antenna is $33 \times 24 \times 0.05\,\text{mm}^3$. The proposed antenna geometry in front and top view is shown in Figure 4.1, and the evolution stages of the proposed antenna are shown in Figure 4.2.

Initially a circular monopole antenna is designed and modified as a ring monopole Ant 2. The slots of square shape slot are made on the top left and bottom right half of the circular ring Ant 3. In Ant 4, the partial ground plane is modified with defects making it a DGS antenna on a plane ring patch. In Ant 5 the ground plane and top patch have square slots. The proposed antenna ground has a larger sized square step slot at the top centre in addition to the two slots in the ground and patch. The radius of the circular ring antenna was calculated using the standard equation (Eq. 4.1) of the UWB antenna at the lower frequency of 3.1 GHz in the UWB band.

$$f = \frac{7.2}{(L + r + p)k} \tag{4.1}$$

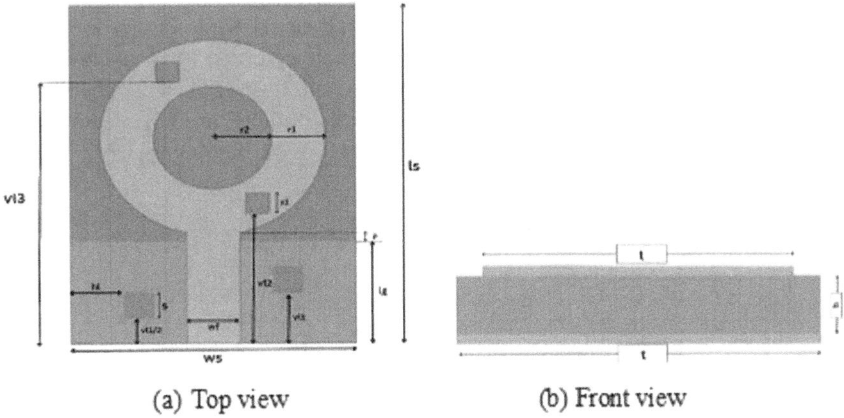


*Figure 4.1* Proposed antenna geometry.

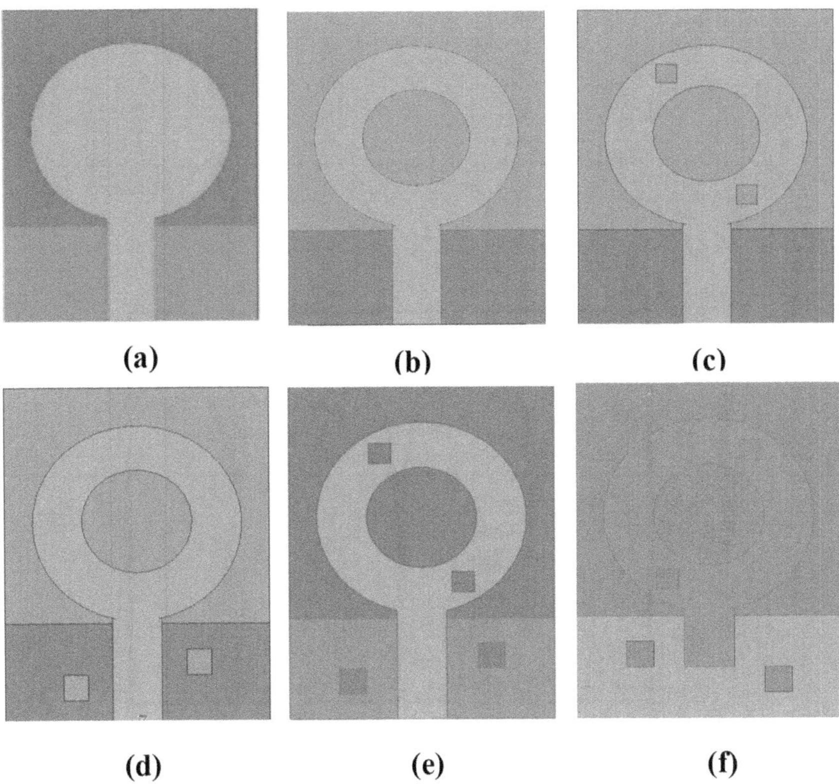


*Figure 4.2* Evolution of proposed antenna (a) Ant 1, (b) Ant 2, (c) Ant 3, (d) Ant 4, (e) Ant 5, … ,(f) proposed.

In Eq. (4.1) $L$ and $r$ are calculated based on the radius of the circle $r1$ as $2r1$ and $r1/4$ respectively. Here $p$ is the spacing of feed length from the ground to the circular monopole. The eccentricity constant $k$ is chosen as 1. Optimal dimensions of the antenna are provided in Table 4.1. The return loss performance graph of all antenna geometry is plotted in Figure 4.3. Proposed antenna geometry has maximum lowest frequency and bandwidth with reference to $S_{11}$<−10 dB.

*Table 4.1* Parameters with dimensions of the antenna

| *Antenna design parameters* | *Dimensions (mm)* | *Antenna design parameters* | *Dimensions (mm)* |
|---|---|---|---|
| H | 2 | $s_1$ | 2 |
| Lg | 10 | *t* | 0.05 |
| Ls | 33 | *hl* | 5 |
| P | 0.5 | $vl_1$ | 5 |
| $r_1$ | 9.5 | $vl_2$ | 12.6 |
| $r_2$ | 5 | $vl_3$ | 25.6 |
| S | 2.5 | *wf* | 4.48 |
| $l_1$ | 5 | ws | 24 |
| $l_2$ | 4.48 | $h_1$ | 12 |

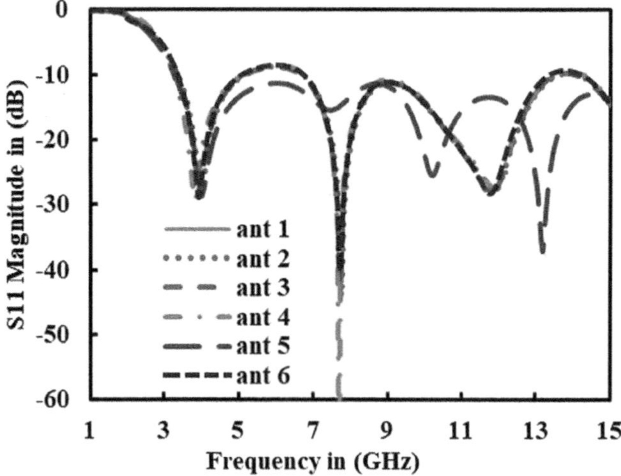


*Figure 4.3* Return loss performance of the antennas during evolution stages.

## 4.3 ANTENNA DESIGN PARAMETER OPTIMIZATION

The structure of the antenna is optimized with various parameters to obtain maximum bandwidth in the UWB band and good return loss. Figure 4.4 shows $S_{11}$ of the proposed antenna for various radii of the outer circle value of $r1$. The radius is varied from 9 to 10 mm in a step of 0.2 mm. When $r1$ = 9.5 mm, it has a maximum bandwidth of 10.1 GHz ranging from 3.35 to 13.45 GHz. Figure 4.5 shows $S_{11}$ of the designed antenna for distinct

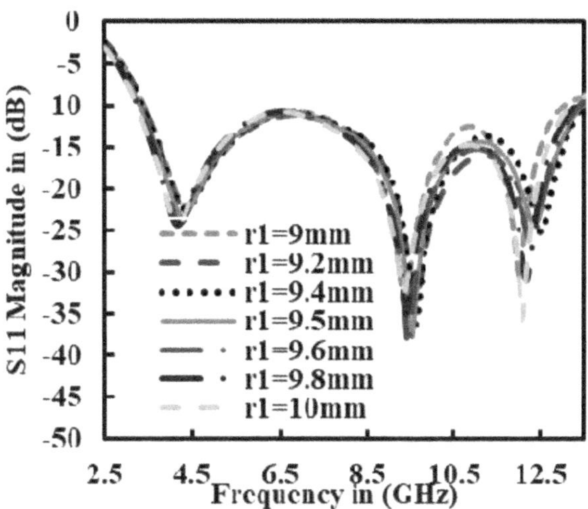


*Figure 4.4* Optimization of outer circle radius *r*1.

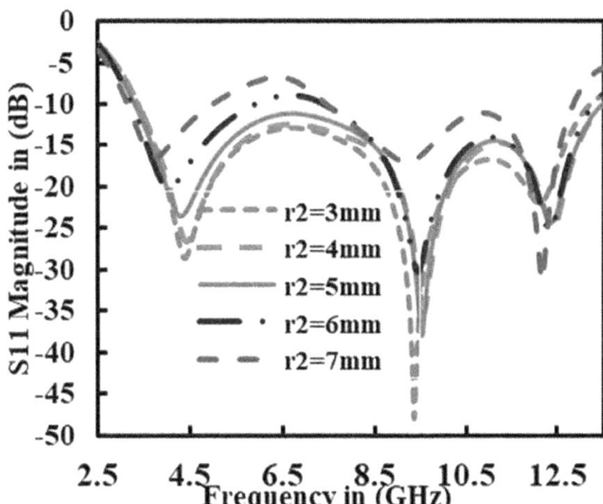


*Figure 4.5* Optimization of inner circle radius *r*2.

values of the radius of inner circle *r2*. Varying *r2* shows a right shift in lower frequency. When *r2*=5 mm, it is found that the magnitude and frequency values are within the UWB band with magnitude reference to –10 dB level.

Figure 4.6 shows $S_{11}$ performance of the proposed antenna for various feed width *wf* and is adjusted to obtain the desired UWB frequencies. It is observed that *wf*=4.48 mm has acquired a good return loss and maximum bandwidth. The lower band is fixed and the upper band frequency reduces with an increase in *wf*. The graph shows a tendency to lower the resonating lowest frequency with an increase in feed width. Figure 4.7 shows $S_{11}$ of the proposed antenna's width of the substrate *ws*. As the width of the substrate hikes, the bandwidth of the antenna reduces, *ws* = 22 mm has maximum bandwidth and return loss, but the width must be carefully chosen to ensure that the impedance matching is optimized for the desired frequency range. Thus, *ws* = 24 mm is chosen to have good radiation characteristics. Figure 4.8 shows $S_{11}$ of the proposed antenna's length of the substrate. The UWB frequency is obtained due to the major frequency shift by keeping *ls*=33 mm and its return loss is also found to be satisfied with good bandwidth. The length of two square slots *s*1 on the radiator is optimized for its dimension and position from the lower ground level. Figure 4.9 shows the slot size variation, while the optimization of slot location is shown in Figures 4.10 and 4.11. The side of the square slot is changed from 1 to 3 mm in intervals of 0.5 mm. A slot with 2 mm length shows maximum bandwidth with $S_{11}$ reference of –10 dB.

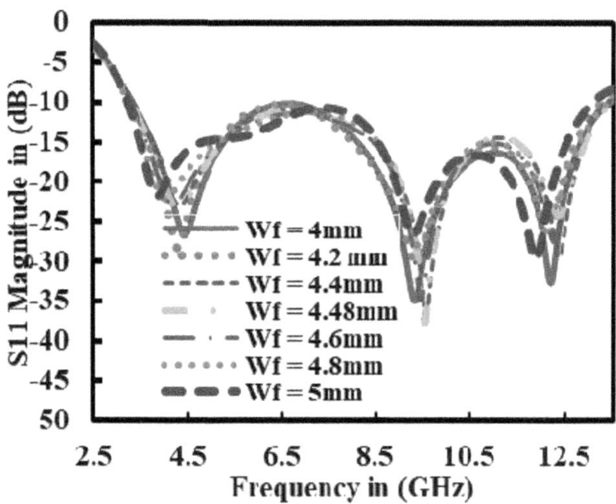


*Figure 4.6* Optimization of feed width *wf*.

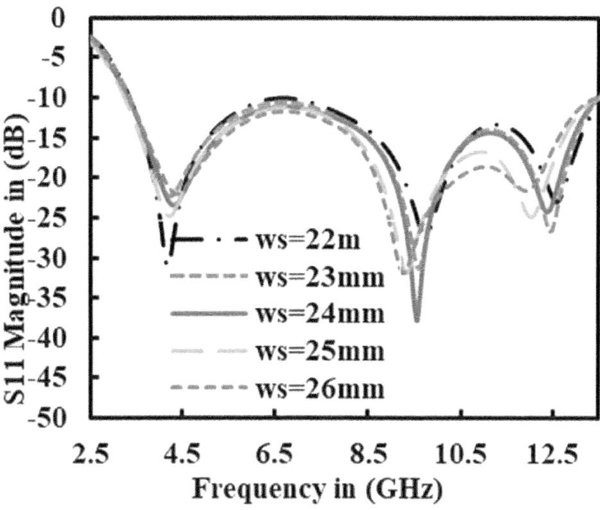


*Figure 4.7* Optimization of substrate width *ws*.

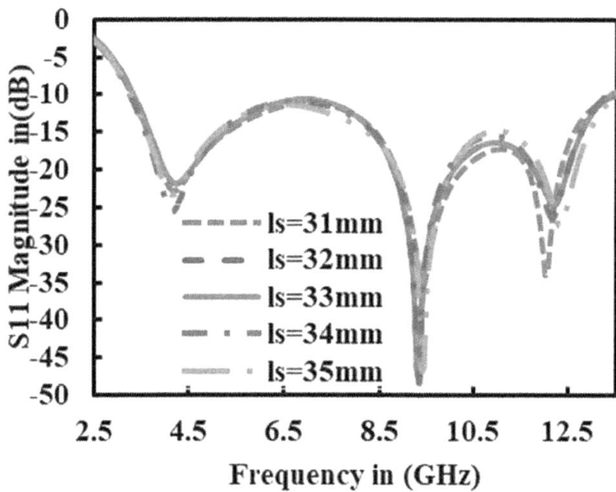


*Figure 4.8* Optimization of substrate length *ls*.

Figure 4.10 shows $S_{11}$ of the proposed antenna which denotes the position of the square slot1 in the circular ring placed at the right half at a distance $vl2$ from the lower ground edge, while keeping $vl2 = 12.6$ mm, the lower resonant frequency has minimum frequency shift with minimal variations

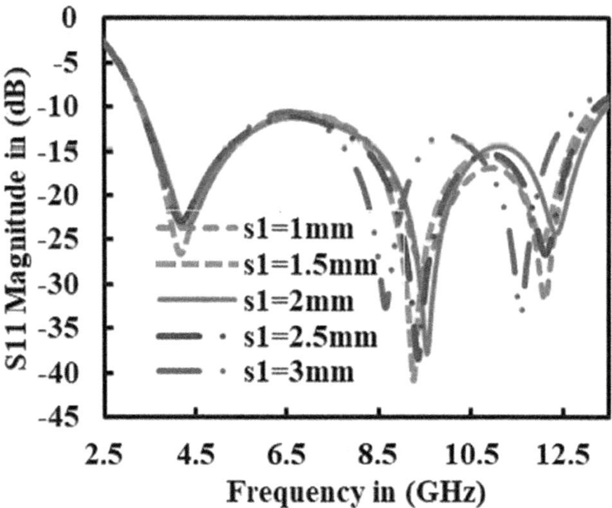


*Figure 4.9* Optimization of radiator slotl sl.

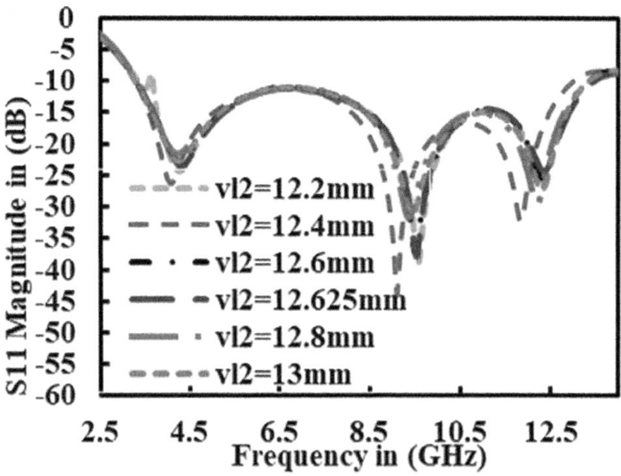


*Figure 4.10* Optimization of slot location *vl2*.

in return loss. In Figure 4.11 the location of the slot in circle on the top left half shows that better response is achieved at $vl3 = 25.6$ mm.

In the ground plane length of the ground *lg*, the square slots side *s*, their location and the centre step size are optimized and plotted in Figures 4.12–4.17. Comparing the performance of various graphs, considering lower

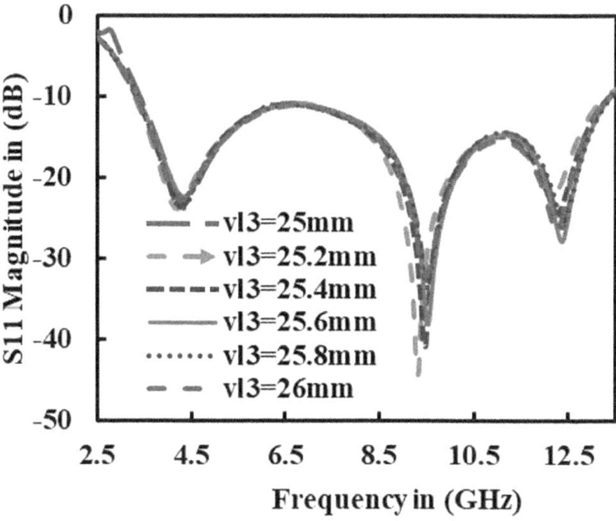


*Figure 4.11* Optimization of slot location *vl*3.

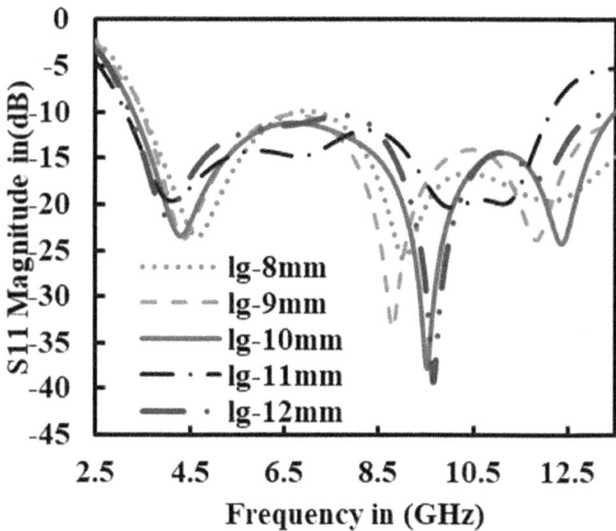


*Figure 4.12* Optimization of ground length *lg*.

frequency and increased bandwidth ground length $lg = 10\,\text{mm}$, square side $s = 2.5\,\text{mm}$, length and width of step cut $l_1$ and $l_2$ as 5 and 4.48 mm and the vertical location of square slots in right as 5 mm and for left slot its half are chosen for best performance. The horizontal location is optimized as 5 mm.

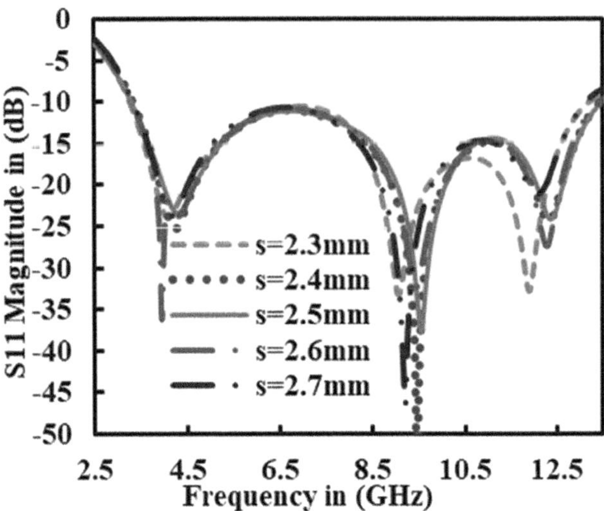


*Figure 4.13* Optimization of ground slot length s.

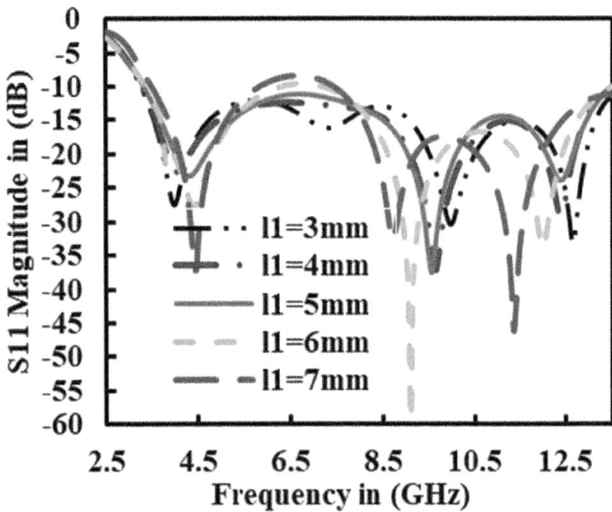


*Figure 4.14* Optimization of ground step vertical height*l*1.

The antenna flexibility is analysed by bending the antenna along X and Y directions. Bending analysis is done by modelling a cylinder of vacuum at a particular radius ranging from 50 to 100 mm whose axis is in the X or Y direction. Ground, substrate and radiator are wrapped on it to make the

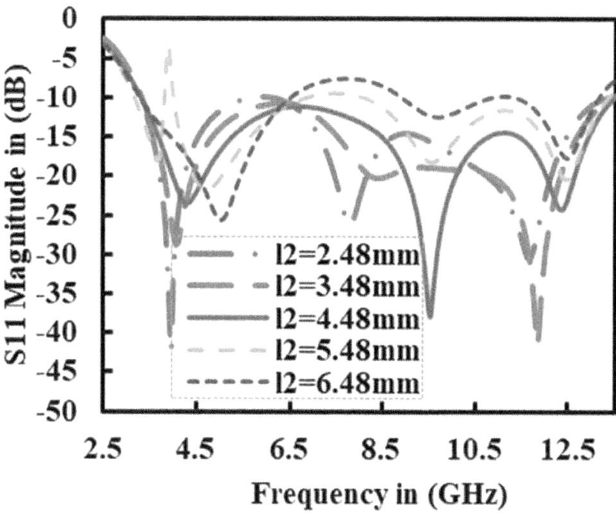


*Figure 4.15* Optimization of ground step horizontal width *l2*.

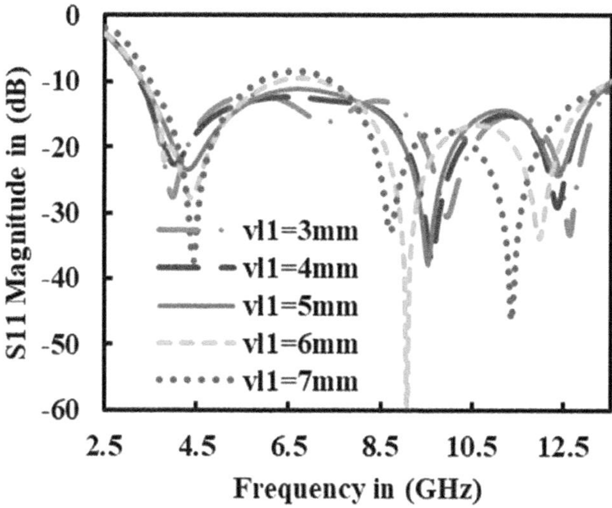


*Figure 4.16* Optimization of vertical length *vl1*

structure bend keeping the antenna centre to cylinder centre. Figures 4.18 and 4.19 show the return loss magnitude results after bending in the X and Y direction, respectively. The results show that the bending radius does not affect the performance of the antenna either in the X or Y direction.

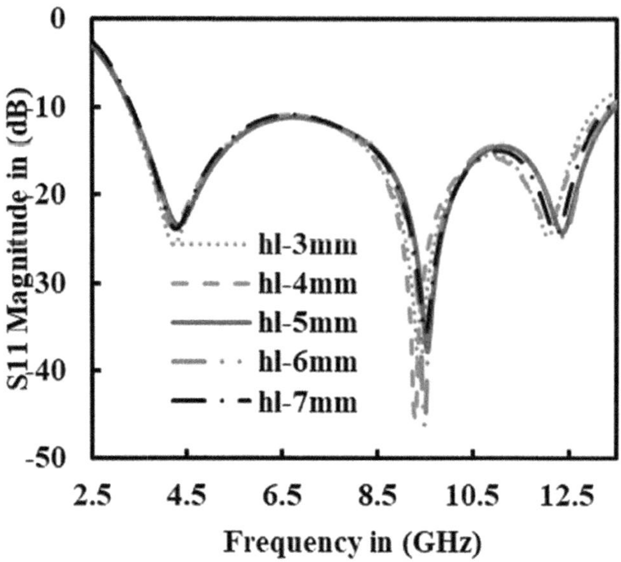


*Figure 4.17* Optimization of horizontal length *hl*.

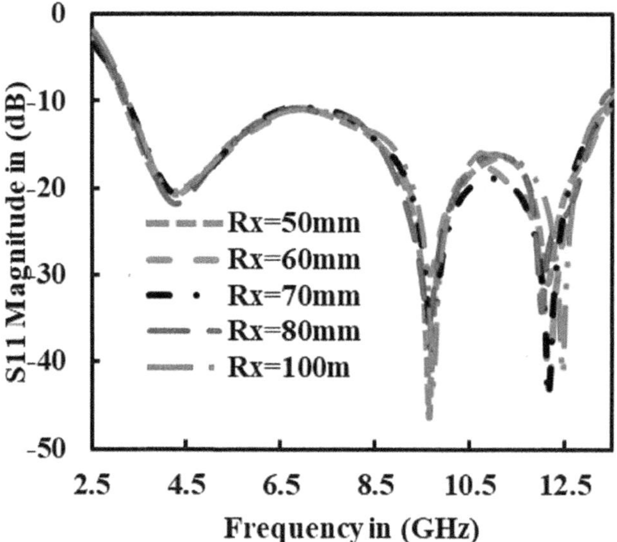


*Figure 4.18* Optimization of X bending.

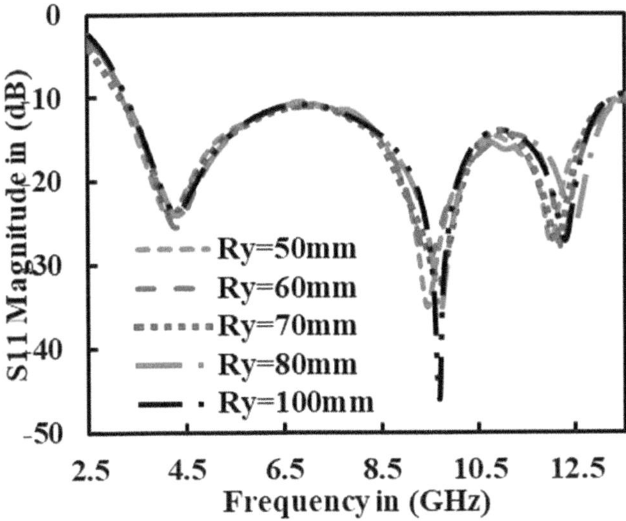


*Figure 4.19* Optimization of Y bending.

## 4.4 BRAIN TUMOUR DETECTION

The proposed antenna is analysed for brain tumour detection by creating a multilayer brain model as shown in Figure 4.20. The structure of the brain phantom is complex and is made up of different tissues. Vast research has been conducted to find out these electrical properties of brain phantoms and how they influence the results, specifically at microwave frequency. In this proposed work, the brain phantom consists of six layers: skin, fat, CSF, brain, dura and bone whose electrical specifications in terms of conductivity, permittivity and loss tangent at lower edge frequency are tabulated in Table 4.2.

The tumour is specified with a sphere of radius 5 mm and dielectric permittivity 71.74. To test the presence of a tumour the antenna is placed near the phantom model with and without a tumour and its performance is analysed. The change in the reflection coefficient of the antenna with and without a tumour, as shown in Figure 4.21, indicates the presence of a tumour.

## 4.5 SAR ANALYSIS

Radiation enters the human body as a result of electromagnetic waves released by antennas. SAR is the absorbed electromagnetic power in W/kg, the amount of energy required by a unit mass of human tissue.

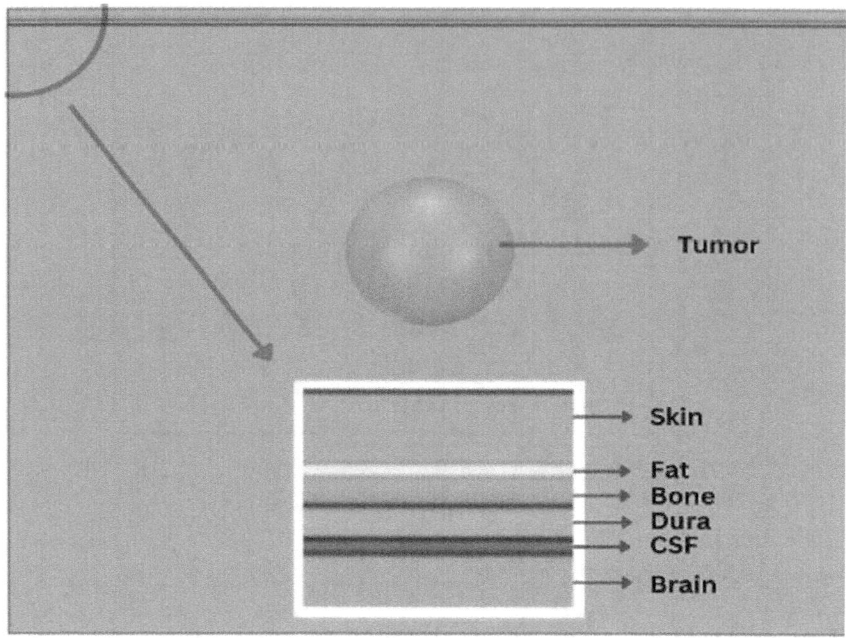


*Figure 4.20* Brain phantom with tumour.

*Table 4.2* Specifications of brain phantom

| *Tissue name* | *Conductivity σ* | *Permittivity εr* | *Loss tangent δ* |
|---|---|---|---|
| Skin | 2.0168 | 41.982 | 0.27855 |
| Fat | 0.13497 | 5.2138 | 0.15011 |
| Bone | 0.52746 | 11.01 | 0.27779 |
| Dura | 2.0824 | 41.215 | 0.29298 |
| CSF | 4.1136 | 65.231 | 0.36567 |
| Brain grey matter | 2.2992 | 47.898 | 0.27834 |
| Brain white matter | 1.5682 | 35.432 | 0.25664 |

SAR limit rules vary amongst international organizations. The Federal Communication Commission standardized the limit for SAR as 1.6 and 2 W/kg for 1 and 10 g of tissue, respectively. Wearable antennas must satisfy these safety requirements. The SAR simulation of the proposed antenna shows a higher SAR value of 7.086 W/kg when placed at a distance of 2 mm from the modelled head phantom. In order to reduce the SAR a Frequency Selective Surface (FSS) backing is provided for the antenna. The FSS is deployed at a distance of $h_1$ just below the ground plane in order to reflect

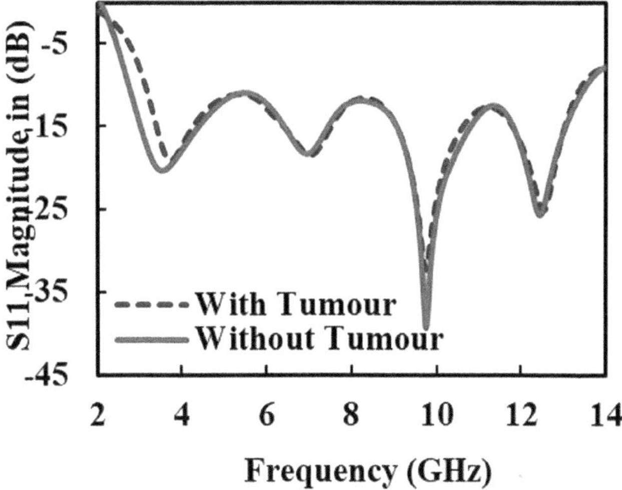


*Figure 4.21* Comparative reflection response of antenna with and without tumour.

the backward radiations, and this placement is essential for achieving the SAR of the proposed antenna.

## 4.6 ANTENNA WITH FSS

The designed antenna has radiation in all directions, which makes the SAR higher. If the back radiation is prevented SAR of the antenna can be reduced. A single-layer UWB FSS proposed by one of the authors operating from 2.39 to 13.67 GHz is used here to reduce the SAR. The FSS is deployed at a distance of $h1$ just below the ground plane in order to reflect the backward radiations, and this placement is essential for achieving the SAR of the proposed antenna. The designed FSS is built on a single side FR-4 substrate with a permittivity of 4.4 and a height of 1.6 mm. The loss tangent for FR-4 is 0.025. The copper conductor used is of thickness 0.035 mm. The structure of UWB FSS consists of loops, dipoles and strips as shown in Figure 4.22 [18]. The FSS array consists of 3×3 unit cell elements to cover the antenna. The unit cell dimensions are listed in Table 4.3. The total size of the FSS array is 52.2×52.2. The height between the antenna and the FSS cell is optimized for higher gain. Height $h1$ indicating the spacing between the antenna and FSS is optimized to be 12 mm and the return loss plots of the FSS antenna are shown in Figure 4.23.

The antenna SAR performance is analysed with various distances from the modelled six layer head phantom. When the antenna is kept at 2 and

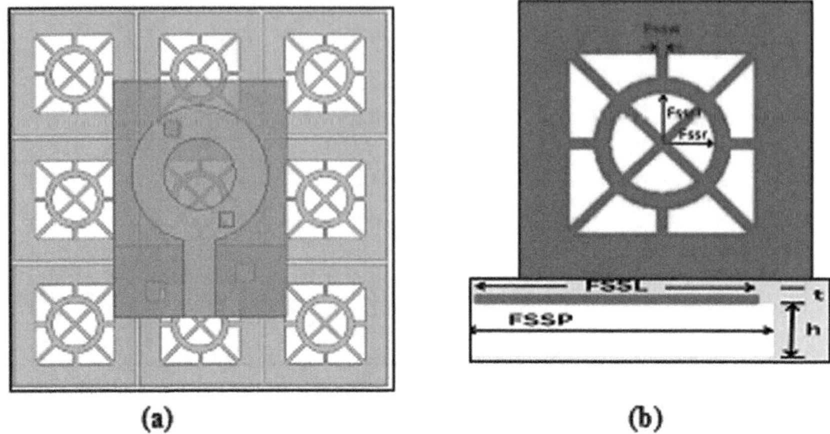


*Figure 4.22* FSS antenna (a) 3×3 array FSS antenna (b) unit cell FSS.

*Table 4.3* Unit cell measurement of the UWB FSS in mm

| *FSSP* | *FSSH* | *FSSL* | *FSST* | *FSSS* | *FSSR* | *FSSR1* | *FSSW* |
|---|---|---|---|---|---|---|---|
| 17.4 | 1.6 | 17 | 0.035 | 11 | 4 | 3.1 | 0.5 |

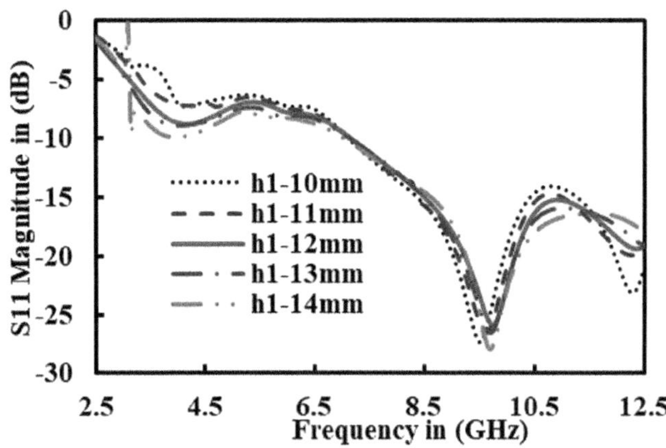


*Figure 4.23* Optimization of FSS antenna height $h_1$.

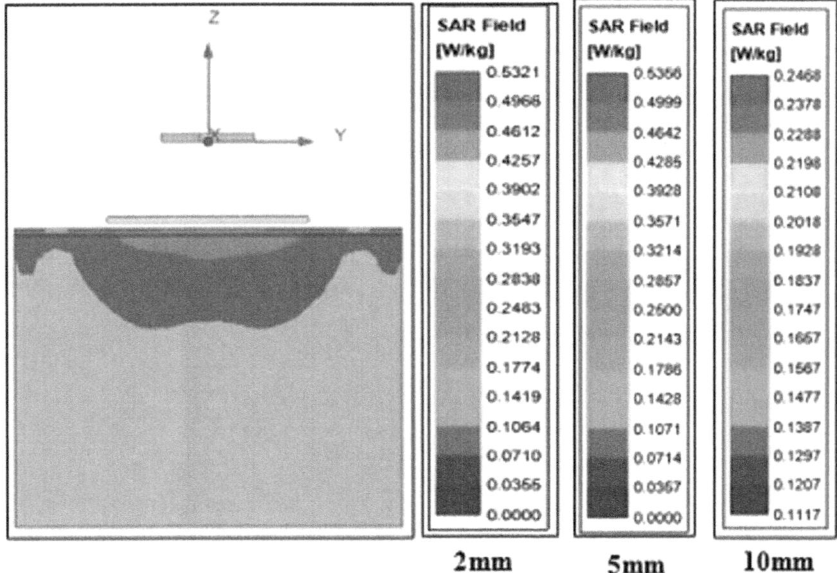


*Figure 4.24* SAR distribution of FSS antenna.

5 mm from the head phantom it is observed that there are not many variations in the obtained SAR whose values are 0.532 and 0.536 W/kg, respectively, as shown in Figure 4.24. At 10 mm the SAR is reduced to half to 0.246 W/kg. Therefore, 10 mm distance is the most suitable for placing the antenna for detecting brain tumours.

The antenna is also checked with a human head male phantom model using HFSS with (WT) and without tumour (WoT), as shown in Figure 4.25. The SAR value of the antenna with and without tumour is reduced using the head phantom compared to the six-layer model. Simulated SAR values for 1 g of tissue are 0.197 and 0.192 W/kg for the head phantom model without and with tumour. The surface current distribution is also analysed for head phantom without and with Tumour and the recorded values are 23.7 and 34.002 A/m. The difference in current distribution can also be taken for determining the tumour. The current distribution also indicates the field distribution of the antenna. The presence of a tumour results in a higher current distribution, which invariably increases the thermal behaviour compared to a head without a tumour. Thus temperature, current distribution or return loss variation can be used to identify the presence of a tumour. These methods can also be applied to microwave imaging.

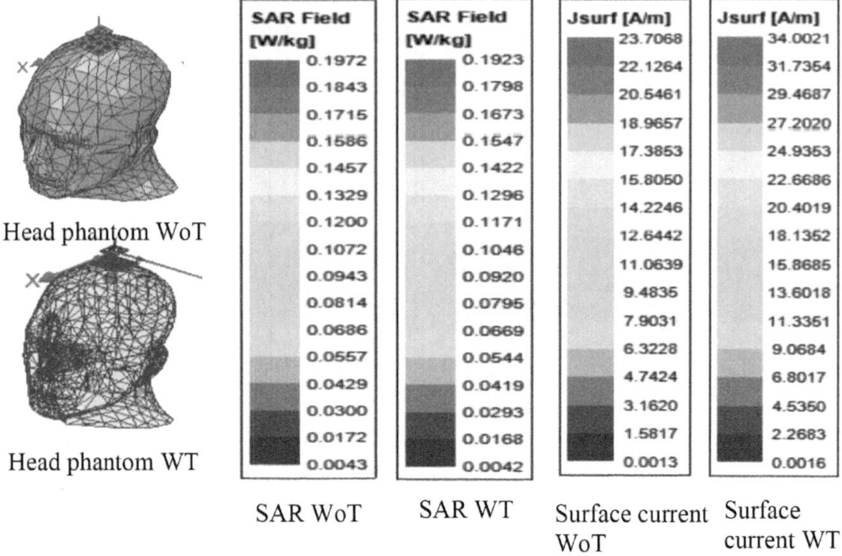


*Figure 4.25* Antenna on human head phantom model indicating SAR and surface current distribution of proposed FSS antenna.

## 4.7 ANTENNA RESULTS AND DISCUSSION

The antenna is fabricated with a dimension of $33 \times 24\,mm^2$ on a 1mm silicon rubber substrate, and the photographs of the prototype are portrayed in Figure 4.26. The fabricated antenna prototype is tested in a specially designed zero reflection zone anechoic chamber measuring $5 \times 3 \times 2.6\,m^3$ against a rigid horn operating from 800 MHz to 18 GHz. Figure 4.27 shows the anechoic chamber with an antenna placed on the mount during measurement. Return loss is tested using a vector network analyser, while gain and radiation patterns are plotted on the system connected to the antenna system in the chamber.

The reflection characteristics of the proposed UWB antenna after fabrication are compared with the simulated return loss in Figure 4.28. The simulated return loss bandwidth of the antenna is 10.1 GHz, with the lower band at 3.35 GHz and the upper band at 13.45 GHz. However, during measurement, the bandwidth was reduced to 7.14 GHz, ranging from 3.87 to 10.94 GHz with a reference of –10 dB. Gain and radiation plots are shown in Figures 4.29 and 4.30, respectively. Peak gain during simulation is 12.4 dBi and during measurement, it is 12.7 dBi. Elevation and azimuth plane radiation patterns at sample frequencies of 3.3 and 9 GHz show a close relation between simulated and measured results.

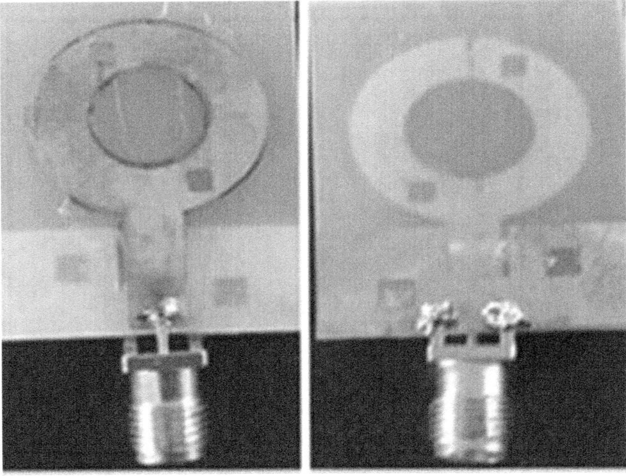

*Figure 4.26* Front and rear view of fabricated antenna.

*Figure 4.27* Antenna testing in an anechoic chamber.

## 4.8 CONCLUSION

In this chapter a flexible ultra-wideband wearable antenna is designed for biomedical application. The fabricated circular ring prototype uses silicon rubber as the dielectric with a defected ground structure and an FSS array of $3 \times 3$ elements. Its performance is good in the UWB band with a maximum

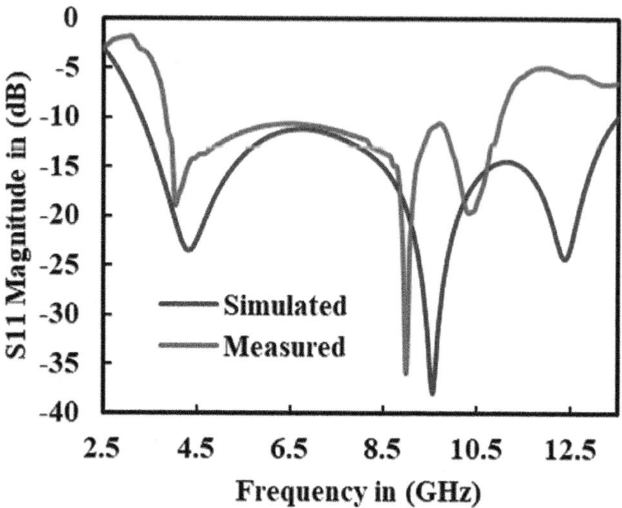


*Figure 4.28* Comparative reflection performance of the designed UWB antenna.

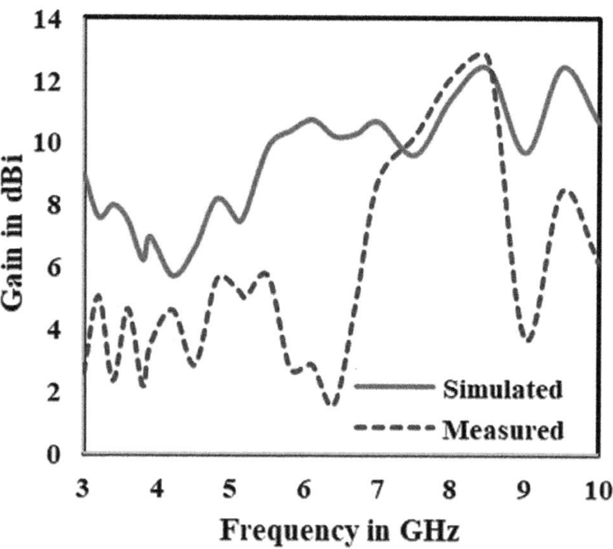


*Figure 4.29* Comparative gain performance of the designed FSS UWB antenna.

bandwidth of 10.1 GHz and an average gain of 9.11 dBi. The antenna also shows a low SAR value of 0.197 W/kg using the head phantom model. The antenna is demonstrated to identify the presence of a brain tumour.

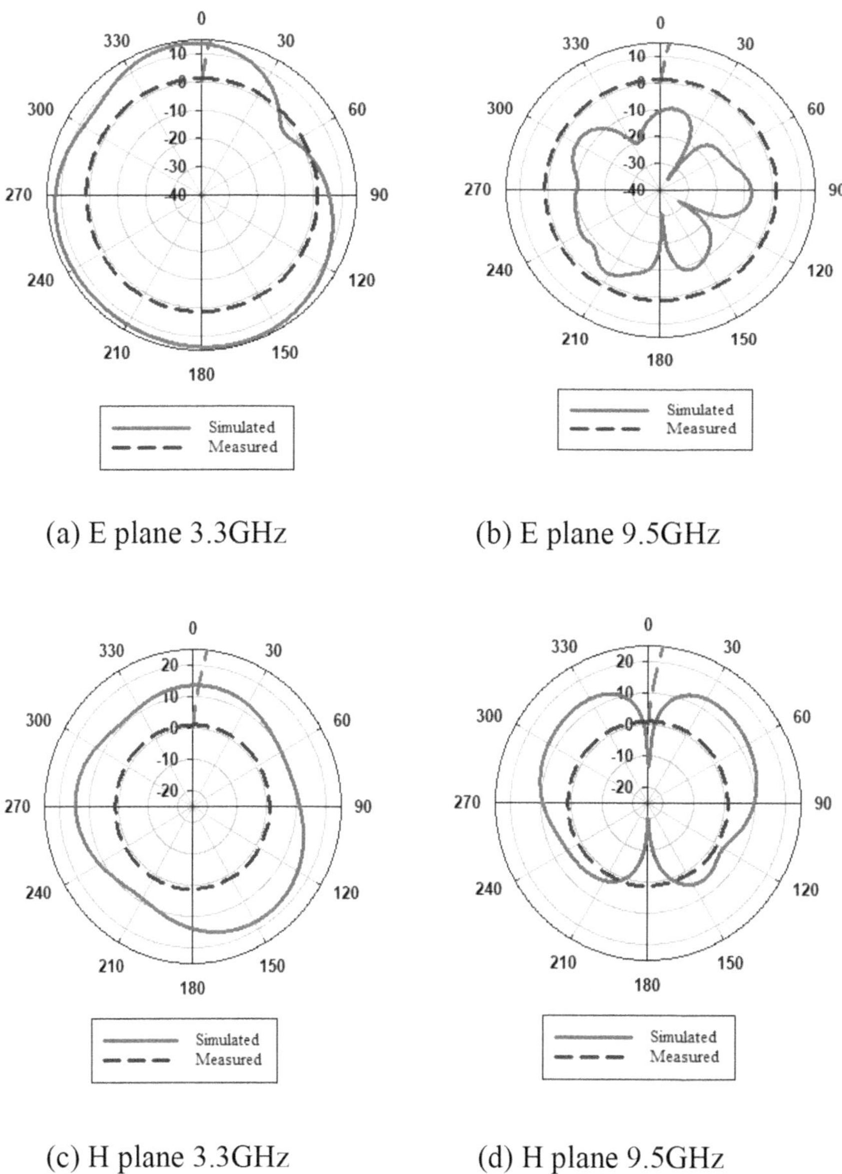


*Figure 4.30* E and H plane radiation patterns of the antenna at 3.3 and 9.5 GHz.

## ACKNOWLEDGEMENT

This work is funded by All India Council for Technical Education, under the research promotion scheme. File No.8-122/FDC/ RPS/POLICY-1/2021-2022.

## REFERENCES


1. J. Amin, M. Sharif, M. Yasmin, S. L. Fernandes, A distinctive approach in brain tumor detection and classification using MRI, *Pattern Recognition Letters*, vol. 139, pp. 118–127, 2020, ISSN 0167-8655, doi: 10.1016/j.patrec.2017.10.036.
2. F. Fahmi, F. Apriyulida, I. K. Nasution, Sawaluddin. Automatic detection of brain tumor on computed tomography images for patients in the intensive care unit, *Journal of Healthcare Engineering*, 2483285, 2020. doi: 10.1155/2020/2483285.
3. B. E. Lyons, R. H. Britt and J. W. Strohbehn, Localized hyperthermia in the treatment of malignant brain tumors using an interstitial microwave antenna array, *IEEE Transactions on Biomedical Engineering*, vol. BME-31, no. 1, pp. 53–62, Jan. 1984, doi: 10.1109/TBME.1984.325370.
4. J. C. Lin, S. Hirai, C.-L. Chiang, W.-L. Hsu, J.-L. Su and Y.-J. Wang, Computer simulation and experimental studies of SAR distributions of interstitial arrays of sleeved-slot microwave antennas for hyperthermia treatment of brain tumors, *IEEE Transactions on Microwave Theory and Techniques*, vol. 48, no. 11, pp. 2191–2198, Nov. 2000, doi: 10.1109/22.884214.
5. A. Hossain et al., A YOLOv3 deep neural network model to detect brain tumor in portable electromagnetic imaging system, *IEEE Access*, vol. 9, pp. 82647–82660, 2021, doi: 10.1109/ACCESS.2021.3086624.
6. Y. A. Hamad, K. Simonov and M. B. Naeem, Brain's tumor edge detection on low contrast medical images, *2018 1st Annual International Conference on Information and Sciences (AiCIS), Fallujah, Iraq*, 2018, pp. 45–50, doi: 10.1109/AiCIS.2018.00021.
7. P. Tiwari, B. Pant, M. M. Elarabawy, M. Abd-Elnaby, N. Mohd, G. Dhiman and S. Sharma, CNN based multiclass brain tumor detection using medical imaging, *Computational Intelligence and Neuroscience*, 1830010, 2022, doi: 10.1155/2022/1830010.
8. J. Amin, M. A. Anjum, M. Sharif, S. Jabeen, S. Kadry and P. M. Ger, A new model for brain tumor detection using ensemble transfer learning and quantum variational classifier, *Computational Intelligence and Neuroscience*, vol. 2022, 13 pages, 2022, doi: 10.1155/2022/3236305.
9. L. C. Paul, M. N. Hossain, M. M. Mowla, M. Z. Mahmud, R. Azim and M. T. Islam, Human brain tumor detection using CPW FED UWB vivaldi antenna, *2019 IEEE International Conference on Biomedical Engineering, Computer and Information Technology for Health (BECITHCON)*, Dhaka, Bangladesh, pp. 1–6, 2019, doi: 10.1109/BECITHCON48839.2019.9063185.
10. S. A. Rezaeieh, A. Zamani and A. M. Abbosh, 3-D wideband antenna for head-imaging system with performance verification in brain tumor detection, *IEEE Antennas and Wireless Propagation Letters*, vol. 14, pp. 910–914, 2015, doi: 10.1109/LAWP.2014.2386852.
11. R. Chishti et al., Advances in antenna-based techniques for detection and monitoring of critical chronic diseases: A comprehensive review, *IEEE Access*, vol. 11, pp. 104463–104484, 2023, doi: 10.1109/ACCESS.2023.3316149.

12. A. Hossain, M. T. Islam, G. K. Beng et al., Microwave brain imaging system to detect brain tumor using metamaterial loaded stacked antenna array, *Scientific Reports*, vol. 12, p. 16478, 2022, doi: 10.1038/s41598-022-20944-8.
13. R. Inum, Md. Masud Rana, K. N. Shushama and Md. Anwarul Quader, EBG based microstrip patch antenna for brain tumor detection via scattering parameters in microwave imaging system, *International Journal of Biomedical Imaging*, vol. 2018, 12 pages, 2018, doi: 10.1155/2018/8241438.
14. D. A. Saleeb, R. M. Helmy, N. F. F. Areed, M. Marey, W. M. Abdulkawi and A. S. Elkorany, A technique for the early detection of brain cancer using circularly polarized reconfigurable antenna array, *IEEE Access*, vol. 9, pp. 133786–133794, 2021, doi: 10.1109/ACCESS.2021.3115707.
15. D. Thangarasu, R. R. Thipparaju, S. K. Palaniswamy, M. Kanagasabai, M. G. N. Alsath, D. Potti and S. Kumar, On the design and performance analysis of flexible planar monopole ultra-wideband antennas for wearable wireless applications, *International Journal of Antennas and Propagation*, vol. 2022, 14 pages, 2022, doi: 10.1155/2022/5049173.
16. R. B. V. B. Simorangkir, A. Kiourti and K. P. Esselle, UWB wearable antenna with a full ground plane based on PDMS-embedded conductive fabric, *IEEE Antennas and Wireless Propagation Letters*, vol. 17, no. 3, pp. 493–496, March 2018, doi: 10.1109/LAWP.2018.2797251.
17. P. Chen, D. Wang, L. Liu, L. Wang and Y. Lin, Design of UWB wearable conformal antenna based on jean material, *International Journal of Antennas and Propagation*, vol. 2022, 12 pages, 2022, doi: 10.1155/2022/4886844.
18. S. Abdulkareem and S. Gopalakrishnan, Modelling of a polarization insensitive UWB FSS with band stop response, *Radioengineering*. vol. 30, no. 2, pp. 342–348, 2021, doi: 10.13164/re.2021.0342.

Chapter 5

# Fractal antennas in medicine

## A comprehensive review of design, performance, and medical applications

*T. Nivethitha, A. Kishore Kumar, and P. K. Poonguzhali*

## 5.1 INTRODUCTION TO FRACTAL ANTENNAS

### 5.1.1 Definition and basic principles of fractal antennas

Fractal antennas are a revolutionary type of antenna that utilize the fascinating world of fractals to achieve remarkable properties. Unlike traditional antennas, which are often bulky and limited in frequency range, fractals offer a compact, multi-band, and highly efficient solution. Imagine an antenna that can fit in the palm of your hand yet operate across multiple frequencies used in cellular phones, Wi-Fi, and even satellite communication. That's the power of fractals! The key to a fractal antenna's magic lies in its self-similar design. This means that the basic pattern or "motif" of the antenna is repeated at different scales, creating intricate geometric shapes. Think of it like a snowflake, where each delicate branch mirrors the overall structure. This repetition effectively increases the electrical length of the antenna within a limited space. More "antennae" packed into a smaller area translates to better signal reception and transmission across a wider range of frequencies.

### 5.1.2 Overview of their unique geometric structures

Fractal antennas exhibit unique geometric structures that are characterized by self-similarity, meaning that the same patterns repeat at different scales. These structures contribute to the antennas' ability to achieve miniaturization, multiband operation, and other desirable properties. Here are some examples of common geometric structures used in fractal antennas:

- **Sierpinski Triangle:** A Sierpinski triangle [1,2] is a fractal pattern that starts with an equilateral triangle. In each iteration, smaller equilateral triangles are removed from the original, creating a self-similar pattern. Sierpinski triangle fractals are used as radiating elements in fractal antennas, and the iterations of triangles can be customized for specific frequency bands.

 DOI: 10.1201/9781003560487-5

- **Koch Snowflake:** The Koch snowflake is a geometric shape created by adding progressively smaller equilateral triangles to the sides of an initial triangle. This iterative process results in a snowflake-like pattern with self-similar structures at different levels. In fractal antennas, the Koch snowflake can be used to create complex radiating structures that contribute to multiband operation.
- **Minkowski Island:** The Minkowski island is a fractal pattern characterized by adding smaller squares to the sides of an initial square in each iteration. This process creates a self-similar, space-filling structure. Minkowski island fractals are employed in fractal antennas to achieve compactness and efficient use of space.
- **Hilbert Curve:** The Hilbert curve is a continuous fractal space-filling curve that traverses a square in a way that preserves its self-similarity. In fractal antennas, the Hilbert curve can be utilized to create radiating elements that cover a large area within a confined space.
- **Menger Sponge:** The Menger sponge is a three-dimensional fractal created by dividing a cube into smaller cubes and removing the centers of each face. While primarily used in three-dimensional structures, the Menger sponge concept can inspire the design of fractal antennas with intricate three-dimensional radiating elements. These geometric structures are just a few examples, and designers can adapt and combine them to create custom fractal antenna designs based on specific performance requirements. The iterative nature of fractals allows for a high degree of customization and optimization, making fractal antennas versatile for various applications in wireless communication [3], RFID systems, and other areas where compact and multiband antennas are needed.

## 5.2 HISTORICAL DEVELOPMENT

### 5.2.1 The evolving journey of fractals in telecommunications: from theory to tiny titans

The tale of fractal antennas in telecommunications [4] is a fascinating one, marked by rapid advancements and ever-expanding possibilities. It's a story of how a theoretical concept blossomed into a practical technology, dramatically shaping the future of wireless communication. Let's delve into the key stages of this evolution:

#### *5.2.1.1 The seeds of innovation (late 1980s)*

The seeds of this revolution were sown by mathematician and electrical engineer Nathan Cohen in the late 1980s. Intrigued by the intriguing properties of fractals, he envisioned their potential to create compact, multiband antennas. His initial research laid the groundwork for exploring fractal geometry's application in antenna design.

#### 5.2.1.2 *Early exploration and experimentation (1990s)*

The 1990s saw the initial exploration and experimentation with various fractal geometries for antenna designs. Pioneers like Cohen, along with other researchers, began creating and testing fractal antennas based on shapes like the Sierpinski gasket, Cantor set, and Koch snowflake. These early prototypes demonstrated the viability of the concept, paving the way for further development.

#### 5.2.1.3 *Optimization and refinement (the early 2000s)*

The early 2000s were marked by significant advancements in the optimization and refinement of fractal antenna designs. Research focused on understanding the relationship between fractal geometry and antenna performance, leading to improvements in gain, bandwidth, and radiation patterns [5]. Advanced computational techniques and simulation tools allowed for more precise prediction and optimization of antenna behavior.

#### 5.2.1.4 *Practical applications and commercialization (2010s–Present)*

The past decade has witnessed the rise of practical applications and commercialization of fractal antennas. Their unique advantages, like miniaturization and multiband operation, made them particularly suitable for mobile devices, wearable electronics, and wireless sensor networks. Companies like Fractus Antennas and Antenova started offering commercially available fractal antenna solutions for various applications.

##### 5.2.1.4.1 *Looking Ahead: The Future of Fractals in Telecommunications*

The future of fractal antennas in telecommunications [6] is vibrant and brimming with possibilities. Research is progressing toward:

- **Multi-Functionality:** Integrating other functionalities like sensing and energy harvesting into fractal antennas.
- **Metamaterial Integration:** Combining fractals with metamaterials to achieve even more exotic properties like cloaking and negative refraction.
- **Adaptive and Reconfigurable Designs:** Creating fractal antennas that can dynamically adjust their properties based on the environment and operating conditions [6–8].

## 5.3 FRACTAL ANTENNA DESIGN PARAMETERS

In the design of fractal antennas, several key parameters play a crucial role in determining the antenna's performance and characteristics. These design parameters include fractal dimension, iteration, and scaling. Let's explore each of these parameters:

### 5.3.1 Fractal dimension

The fractal dimension is a measure of the complexity and self-similarity of a fractal shape. In the context of fractal antennas, different fractal dimensions can be employed to achieve specific design goals. A higher fractal dimension often corresponds to increased complexity and more intricate structures. This complexity can contribute to multiband operation and improved radiation characteristics. Common fractal dimensions used in antenna design include values between 1 and 2, as they represent shapes that fill more space or have more complex structures.

### 5.3.2 Iteration

Iteration refers to the number of times a basic fractal element or pattern is repeated in the generation of the final antenna structure. Increasing the number of iterations generally leads to a more detailed and refined fractal shape. Higher iteration levels contribute to finer details and increased complexity, allowing for greater customization and optimization of antenna characteristics. However, there may be practical limitations in terms of fabrication and computational resources, so the choice of the iteration level often involves a trade-off between complexity and feasibility [8].

### 5.3.3 Scaling

Scaling involves adjusting the size of the fractal elements in each iteration. Scaling can affect the electrical characteristics of the antenna, including impedance and resonant frequencies. It allows designers to create antennas that are compact while maintaining efficient radiation properties. The choice of scaling factor is critical, as it influences not only the physical dimensions but also the electrical performance of the antenna.

### 5.3.4 Geometric configurations

Different fractal shapes and configurations contribute to the overall design of the antenna. For example, Sierpinski triangles, Koch curves, Minkowski islands, and Hilbert curves are common fractal geometries used in antenna design [7]. The selection of a specific geometric configuration depends on the desired properties such as multiband operation, miniaturization, or space-filling characteristics.

### 5.3.5 Material considerations

The choice of materials for constructing the fractal antenna also influences its performance. Material properties such as conductivity and dielectric constant can impact the antenna's efficiency and bandwidth. Fractal antennas can be designed to operate on various substrates, and material choices are often made based on the specific application requirements.

### 5.3.6 Frequency range

The design parameters need to be carefully tuned to cover the desired frequency range. Fractal antennas are versatile and can be customized for specific bands or designed to operate across a wide frequency spectrum.

### 5.3.7 Practical constraints

Design parameters must consider practical constraints, including fabrication techniques, manufacturing tolerances, and the capabilities of the intended application or device. By manipulating these design parameters, antenna engineers and researchers can tailor fractal antennas to meet specific performance criteria, making them suitable for a wide range of applications in telecommunications, wireless communication, and other fields. The iterative and customizable nature of fractal antenna design allows for flexibility in achieving desired electrical and physical characteristics.

### 5.3.8 Influence of fractal geometry on antenna performance

Fractal geometry has a profound influence on antenna performance, and its incorporation into antenna design has led to several notable improvements and advantages. Here are some key aspects of how fractal geometry affects antenna performance:

**Multiband Operation**: One of the significant benefits of fractal antennas is their ability to operate across multiple frequency bands [9]. The self-similar and space-filling properties of fractals allow for effective coupling at different scales, enabling antennas to cover a broad frequency range.

**Miniaturization**: Fractal antennas are known for their ability to achieve miniaturization without sacrificing performance. The intricate and self-repeating patterns of fractals allow for a compact design with a reduced physical footprint while maintaining efficient radiation properties.

**Increased Aperture and Effective Length:** Fractal geometry can effectively increase the effective electrical length of an antenna, which is crucial for achieving resonance at lower frequencies. This property is beneficial for antennas operating in the lower frequency bands [10] where longer physical lengths are traditionally required.

**Improved Radiation Patterns:** The complexity and self-similarity of fractal patterns contribute to improved radiation patterns. Fractal antennas can exhibit enhanced directivity, gain, and radiation efficiency compared to conventional antenna designs, leading to better overall performance.

### 5.3.9 Space-filling characteristics

Fractals often exhibit space-filling properties, allowing the efficient use of available space for radiating elements. This property is advantageous in situations where a larger radiating area is needed within a confined space.

### 5.3.10 Frequency agility

Fractal antennas offer frequency agility, meaning they can be designed to operate at various frequencies within their bandwidth. This flexibility is crucial in applications where the operating frequency may change or where multiband operation is required.

### 5.3.11 Resilience to damage

The self-similar structure of fractal antennas often provides a degree of redundancy. Even if a portion of the antenna is damaged or compromised, the remaining structure can still exhibit similar performance characteristics, contributing to the antenna's resilience.

### 5.3.12 Customization and optimization

Fractal geometry allows for a high degree of customization and optimization in antenna design. Engineers can tailor the fractal patterns, iterations, and scaling factors to achieve specific performance goals, making fractal antennas adaptable to various applications.

### 5.3.13 Compact and conformal designs

Fractal antennas can be designed in compact and conformal shapes, making them suitable for integration into a wide range of devices and structures. Their flexibility in design enables adaptation to non-traditional antenna shapes and forms.

### 5.3.14 Enhanced bandwidth

Fractal antennas often exhibit wide bandwidth characteristics. The self-similar structures and iterative design contribute to the antenna's ability to cover a broad frequency range, making them suitable for applications where bandwidth is a critical factor.

### 5.3.15 Efficient use of available space

Fractal antennas are capable of efficiently using available space, making them suitable for applications with spatial constraints. This property is particularly valuable in modern electronic devices where space optimization is crucial. The influence of fractal geometry on antenna performance is evident in the array of advantages it brings, making fractal antennas valuable components in wireless communication systems, RFID technology, and other applications where compact, multiband, and efficient antennas are essential [11].

## 5.4 PERFORMANCE CHARACTERISTICS

The intricate world of fractal antennas not only boasts impressive miniaturization and multiband capabilities but also shines in its **gain and directivity**. These crucial performance characteristics define how efficiently an antenna amplifies the signal it transmits and focuses it in specific directions, making it a true powerhouse in the realm of wireless communication.

### 5.4.1 Gain and directivity of fractal antennas

#### *5.4.1.1 Gain*

Gain is a measure of the antenna's ability to concentrate radiation in a particular direction. It is the ratio of the power radiated in the desired direction to the power radiated isotropically (equally in all directions). Fractal antennas [12] can achieve high gain due to their complex and self-similar structures. The intricate patterns and details at various scales contribute to constructive interference in the desired direction, resulting in increased gain. The miniaturization capability of fractal antennas allows for the creation of compact designs with high gain, making them suitable for applications where space is limited.

The most common equation for characterizing the gain ($G$) of an antenna relative to an isotropic radiator is:

$$G = 4\pi * F(\theta,\varphi)/\Omega \; G=4\pi * \mathrm{F}(\theta, \varphi) /\Omega$$

- $F(\theta,\varphi)$: Radiation pattern function, describing the power density radiated in specific directions ($\theta$, $\varphi$) by the antenna. For fractals, this function can be complex due to their intricate geometries.
- **Ω:** Solid angle representing the entire radiation sphere ($4\pi$ steradians).

To understand how fractals impact gain, let's analyze $F(\theta,\varphi)$:

- **Increased Electrical Length:** Fractals often achieve this through self-similar iterations, effectively "cramming" more antenna into a smaller space. This leads to a larger effective aperture area, potentially increasing radiated power density in certain directions, boosting $F(\theta,\varphi)$.
- **Directional Control:** Carefully tailoring the fractal pattern can focus the radiation pattern in specific directions, further enhancing $F(\theta,\varphi)$ in those desired directions.

#### *5.4.1.2 Directivity*

Directivity is a measure of how well an antenna focuses its radiated energy in a specific direction. It is closely related to gain but takes into account the antenna's three-dimensional radiation pattern. Fractal antennas often exhibit enhanced directivity because of their ability to create complex radiation patterns with sharp beam edges. The self-similar structures contribute to the formation of intricate lobes and nulls in the radiation pattern, allowing for precise control over the directionality.

The space-filling characteristics of fractals also contribute to improved directivity by efficiently utilizing the available space for radiating elements. While gain measures the overall signal amplification, directivity ($D$) focuses on the directional concentration of the radiated power:

$$D = G/G_{\text{iso}}$$

- $G_{\text{iso}}$: Gain of an isotropic radiator (1 dB).

Higher values of directivity indicate a more focused radiation pattern, minimizing energy waste in unwanted directions. For fractals, achieving high directivity often involves:

- **Shaped Radiation Patterns:** The intricate geometry of fractals allows for tailoring the antenna's $F(\theta,\varphi)$ to concentrate power in specific directions, minimizing the spread in other directions and consequently increasing $D$.

### 5.4.2 Performance trade-offs

It's important to remember that optimizing for one characteristic might impact the other:

- **Gain vs. Directivity:** Increasing directivity often involves focusing the radiation pattern in a specific direction, potentially sacrificing some overall gain compared to a broader pattern.
- **Bandwidth vs. Gain/Directivity:** Achieving wider bandwidth might require compromising on the peak gain and directivity values as the antenna's resonance behavior becomes less focused. Understanding these trade-offs is crucial for designing fractal antennas for specific applications and balancing desired performance characteristics with practical constraints.

### 5.4.3 Multiband directivity

Fractal antennas can be designed to exhibit directivity across multiple frequency bands. This multiband capability is advantageous in modern communication systems [13] where the demand for antennas operating in diverse frequency ranges is common. The self-similar and scalable nature of fractal designs allows for the creation of antennas that maintain consistent directivity characteristics across different frequency bands.

### 5.4.4 Efficiency and bandwidth

The efficiency of an antenna, which is its ability to convert input power into radiated power, is crucial for overall performance. Fractal antennas often exhibit high efficiency due to their optimized design, contributing to better gain and directivity. Fractal antennas are known for their wide bandwidth characteristics. This allows them to maintain good performance over a broad frequency range, making them suitable for applications where bandwidth is a critical consideration.

**Radiation Pattern Control:** Fractal antennas provide engineers with a high degree of control over the radiation pattern. Through careful design of fractal elements, iterations, and scaling, engineers can shape the radiation pattern to meet specific requirements, such as achieving focused beams or covering specific sectors.

### 5.4.5 Resilience to environmental effects

The self-similar and redundant structures of fractal antennas can contribute to resilience against environmental effects or structural damage.

Even if part of the antenna is compromised, the remaining structure may still exhibit similar gain and directivity characteristics. In summary, the gain and directivity of fractal antennas benefit from the self-similar, space-filling, and complex geometric properties of fractal designs. These characteristics enable the creation of antennas with high performance, making fractal antennas suitable for a wide range of applications in telecommunications [7], satellite communication, wireless networks, and more. The polarization properties and impedance matching are essential aspects of antenna design that influence the performance and efficiency of the antenna system. Fractal antennas exhibit unique characteristics in these aspects, and understanding how fractal geometry affects polarization and impedance matching is crucial. Here's an overview:

### 5.4.6 Polarization properties

#### *5.4.6.1 Polarization types*

Fractal antennas can be designed to exhibit different polarization types, including linear polarization (vertical or horizontal), circular polarization (right-hand or left-hand), and elliptical polarization. The fractal geometry allows for flexibility in tailoring the polarization properties based on the application requirements.

#### *5.4.6.2 Multiband polarization*

Fractal antennas can achieve multiband polarization characteristics. This means that the antenna can exhibit specific polarization properties across different frequency bands, providing adaptability to diverse communication standards.

#### *5.4.6.3 Complex geometries for polarization control*

The intricate and self-similar structures of fractal antennas enable engineers to design complex geometries that contribute to polarization control. Fractal elements can be customized to generate specific polarization states, and the iterative nature of fractal designs allows for fine-tuning polarization properties.

#### *5.4.6.4 Wideband polarization*

Fractal antennas often exhibit wideband characteristics, allowing them to maintain consistent polarization properties over a broad frequency range. This wideband [16] behavior is beneficial in applications where the polarization requirements may vary across different frequency bands.

### 5.4.7 Impedance matching

#### *5.4.7.1 Enhanced bandwidth*

Fractal antennas are known for their ability to operate over wide bandwidths. This wideband behavior contributes to improved impedance matching because the antenna can maintain a low voltage standing wave ratio over a broad frequency range.

### 5.4.8 Iterative design for matching networks

The iterative nature of fractal designs allows engineers to fine-tune the impedance-matching characteristics of the antenna. Matching networks can be integrated into the fractal antenna structure to optimize impedance matching at specific frequencies or across multiple bands.

### 5.4.9 Customizable input impedance

Fractal antennas offer a high degree of customization in terms of input impedance. By adjusting the fractal geometry and scaling factors, designers can tailor the input impedance to match the impedance of the feeding transmission line or system.

### 5.4.10 Minimization of reflections

The self-similar and space-filling properties of fractal antennas contribute to minimizing reflections within the antenna structure. This helps in achieving better impedance matching and efficient power transfer between the antenna and the connected system.

### 5.4.11 Adaptation to different feed configurations

Fractal antennas can be designed to adapt to various feed configurations, allowing for impedance matching with different types of feed networks and transmission lines.

### 5.4.12 Biocompatibility considerations

- **Material Selection:** The materials used to fabricate the antenna must be biocompatible, non-toxic, and resistant to corrosion within the human body. Polymers, ceramics, and certain metals can be suitable options based on the specific application.
- **Tissue Interaction:** The antenna's electrical field distribution and power density need to be carefully controlled to avoid tissue heating, nerve stimulation, or interference with vital functions.

- **Long-term Safety:** The biocompatibility of the antenna materials and design should be assessed for long-term stability and potential degradation within the body.
- **Regulatory Requirements:** Medical devices incorporating any antenna technology need to comply with stringent regulations regarding biocompatibility and safety before clinical use.

### 5.4.13 Safety considerations in the context of human exposure to electromagnetic fields

Safety considerations regarding human exposure to electromagnetic fields (EMFs) [14] are crucial when designing and deploying fractal antennas, especially in applications where the antennas are in close proximity to individuals. The potential health effects associated with EMF exposure have been a subject of study, and regulations and guidelines have been established to ensure safety. Here are some safety considerations in the context of human exposure to EMFs from fractal antennas:

### 5.4.14 Specific absorption rate (SAR) limitations

SAR is a measure of the rate at which energy is absorbed by the body when exposed to EMFs. Fractal antenna designs should be assessed to ensure that SAR levels comply with established safety limits. The antenna's proximity to the human body and its radiation characteristics should be considered in SAR calculations.

### 5.4.15 Frequency and wavelength considerations

The frequency of operation is a critical factor in determining the potential health effects of EMF exposure. Different frequencies penetrate the body to varying depths. Fractal antennas designed for medical applications or wearable devices, for example, should operate at frequencies that minimize penetration into sensitive tissues.

### 5.4.16 Distance and power density

Maintaining a safe distance between the antenna and individuals is an effective way to reduce exposure. Power density, which is the power per unit area, decreases with distance from the source. Fractal antenna [15] installations should consider the appropriate distance to ensure that power density levels conform to safety guidelines.

### 5.4.17 Time-averaging and duty cycle

Safety standards often consider time-averaging over a specific period. Fractal antennas should be designed and operated with consideration for time-averaging to prevent excessive exposure during prolonged periods. Duty cycle, or the proportion of time the antenna is actively transmitting, should also be taken into account.

### 5.4.18 Children and vulnerable populations

Children and certain populations may be more vulnerable to the effects of EMF exposure. Extra precautions may be necessary to protect these groups. Fractal antennas used in applications where children or vulnerable populations may be exposed should undergo additional safety assessments.

### 5.4.19 Occupational exposure

In occupational settings where individuals may be near fractal antennas, safety measures should be implemented to minimize exposure. Training and awareness programs can educate personnel on the potential risks associated with EMF exposure and the necessary precautions.

### 5.4.20 Labeling and warning signs

Clearly labeling devices or areas with appropriate warning signs can help inform individuals about potential EMF exposure. Users and personnel should be aware of safety precautions and guidelines associated with the specific fractal antenna application.

## 5.5 MEDICAL IMAGING APPLICATIONS

### 5.5.1 Use of fractal antennas in medical imaging systems

Fractal antennas find applications in medical imaging systems, contributing to advancements in imaging technology and providing solutions to specific challenges. Their unique characteristics, such as compact size, multiband functionality, and wide frequency range coverage, enable their application across diverse imaging modalities. Here are a few applications of fractal antennas in medical imaging systems.

### 5.5.2 Magnetic resonance imaging (MRI)

Fractal antennas can be employed in the design of RF coils used in MRI systems. These RF coils are crucial for transmitting RF pulses into the body and receiving the resulting signals. The compact size and multiband capabilities of fractal antennas make them suitable for constructing RF coils that can operate efficiently across different imaging sequences and anatomical regions.

### 5.5.3 Computed tomography (CT)

In CT imaging, antennas are used in detectors to capture X-ray signals. Fractal antennas can be designed to enhance the sensitivity and efficiency of X-ray detectors. The space-filling characteristics of fractals may contribute to optimizing the layout of detectors, ensuring efficient coverage and improved image quality.

### 5.5.4 Ultrasound imaging

Fractal antennas can be used in ultrasonic transducers for medical ultrasound imaging applications. Their compact size and multiband capabilities allow for versatile transducer designs. The self-similar patterns of fractal antennas can be advantageous in creating transducer arrays with efficient and customizable acoustic characteristics.

### 5.5.5 Microwave imaging

Fractal antennas operating in the microwave frequency range can be utilized in emerging medical imaging techniques such as microwave imaging and microwave tomography. The complex and self-similar structures of fractal antennas can contribute to creating antennas with unique radiation patterns, facilitating specific imaging requirements.

### 5.5.6 Functional near-infrared spectroscopy (fNIRS)

Fractal antennas can be incorporated into optical systems used in functional near-infrared spectroscopy. In fNIRS, near-infrared light is used to measure changes in blood oxygenation levels in the brain. Fractal antennas may contribute to the design of compact and efficient optical components, enhancing the performance of the imaging system.

### 5.5.7 Intravascular imaging

Fractal antennas may be utilized in intravascular imaging systems, where miniaturized antennas are needed for imaging within blood vessels. The small size and multiband operation of fractal antennas can be advantageous for creating probes or catheters with imaging capabilities for intravascular applications.

### 5.5.8 Wireless capsule endoscopy

Fractal antennas can be employed in wireless capsule endoscopy systems for imaging the gastrointestinal tract. The compact and efficient design of fractal antennas can contribute to the miniaturization of the wireless communication components within the endoscopic capsule.

### 5.5.9 Customized antenna arrays

Fractal antennas provide flexibility in designing customized antenna arrays tailored to specific imaging requirements. The ability to optimize antenna characteristics, such as gain and directivity, allows for the creation of arrays that enhance imaging resolution and sensitivity. The use of fractal antennas in medical imaging systems showcases their versatility and adaptability to different imaging modalities. As technology continues to advance, fractal antennas may play a significant role in improving the performance, miniaturization, and efficiency of medical imaging devices, contributing to better diagnostic capabilities and patient care.

## 5.6 WIRELESS CAPSULE ENDOSCOPY

### 5.6.1 Challenges of wireless communication in capsule endoscopy

- **Miniaturization:** Traditional antennas are often bulky and unsuitable for integration into the tiny capsules used for internal examinations.
- **Limited Battery Life:** Capsules rely on small batteries, necessitating efficient communication to minimize power consumption.
- **Tissue Attenuation:** Signals need to penetrate different bodily tissues effectively for clear data transmission.
- **Multipath Interference:** The complex internal environment leads to signal reflections and scattering, potentially distorting communication.

### 5.6.2 Advantages of fractal antennas for capsule endoscopy

- **Miniaturization:** Fractal geometries enable significantly smaller antennas compared to conventional designs, fitting snugly within the capsule's limited space.
- **Multiband Functionality:** A single fractal antenna can resonate at multiple frequencies, facilitating communication with various medical devices operating across different bands.
- **Enhanced Signal Penetration:** Specific fractal geometries can focus and transmit signals through tissue with less energy loss, improving communication reliability.
- **Reduced Antenna Losses:** Fractal designs can offer higher radiation efficiency, minimizing power consumption and extending battery life.

## 5.7 IMPROVING COMMUNICATION AND DATA TRANSFER WITHIN THE HUMAN BODY

The human body presents a unique and challenging environment for wireless communication, posing limitations on traditional antennas due to their size, efficiency, and safety concerns. However, fractal antennas with their intricate and space-efficient designs emerge as potential game-changers, promising improved communication and data transfer within the body. Let's explore this captivating field:

### 5.7.1 Advantages of fractal antennas for body area networks (BANs)

- **Miniaturization:** Fractal geometries enable significantly smaller antennas compared to conventional designs, making them ideal for integration into wearable devices or implantable sensors.
- **Multiband Functionality:** A single fractal antenna can resonate at multiple frequencies, facilitating communication with various medical devices operating across different bands.
- **Conformal Flexibility:** Certain fractal designs can adapt to the curves of the human body, ensuring comfortable integration with sensors or implants and improved signal transmission.
- **Enhanced Tissue Penetration:** Specific fractal geometries can focus and transmit signals through tissue with less energy loss, improving communication reliability and reducing power consumption.
- **Reduced Antenna Losses:** Fractal designs offer higher radiation efficiency, minimizing power consumption and extending the battery life of wearable devices or implanted sensors.

### 5.7.2 Applications of improved communication and data transfer in BANs

- **Real-Time Health Monitoring:** Sensors embedded in clothing or implanted within the body can continuously track vital signs like heart rate, temperature, and blood pressure, enabling real-time data transmission for improved medical monitoring and preventive healthcare.
- **Neurological Monitoring and Treatment:** Fractal antennas can facilitate communication with brain-computer interfaces or deep brain stimulation devices, offering new avenues for treating neurological disorders like Parkinson's or epilepsy.
- **Wound Healing and Rehabilitation:** Sensor-equipped bandages or implants can track healing progress, provide real-time feedback for personalized treatment plans, and monitor potential complications.

- **Drug Delivery and Therapy:** Implantable microfluidic devices utilizing fractal antenna communication can release medication in response to real-time physiological data, enabling personalized and precise drug delivery.
- **Remote Patient Care:** Wireless communication enables continuous monitoring and data collection from patients in remote locations, improving accessibility to healthcare and reducing hospitalizations.

## 5.8 ENABLING WIRELESS COMMUNICATION FOR TELEMEDICINE APPLICATIONS

Fractal antennas play a significant role in remote patient monitoring by enabling wireless communication between wearable medical devices and monitoring systems. Here are some key aspects of their role:

**Compact Design:** Fractal antennas are known for their compact size and efficient use of space. This characteristic is crucial in the design of wearable devices for remote patient monitoring. Fractals allow antennas to be built on small, flexible substrates, making them easy to integrate into wearable devices without adding bulk or discomfort for the patients.

**Wideband Capability:** Fractal antennas have a wideband frequency response, allowing them to operate over a broad range of frequencies simultaneously. This capability is vital in remote patient monitoring applications where multiple signals, such as vital signs data, need to be transmitted wirelessly. Fractals enable the antenna to efficiently receive and transmit signals across various frequency bands, ensuring seamless communication.

**Enhanced Signal Strength and Coverage:** The intricate geometric design of fractal antennas helps improve signal strength and coverage. Their self-similar structure allows for more surface area within a given space, leading to enhanced signal reception and transmission. This is particularly beneficial in remote patient monitoring, where reliable wireless connectivity is crucial for continuous monitoring and timely transmission of data.

**Reduced Interference:** Fractal antennas have inherent filtering capabilities due to their intricate geometries. They can suppress or attenuate unwanted signals from nearby sources, reducing interference and improving signal quality. In remote patient monitoring, this is essential to ensure accurate and reliable transmission of physiological data, free from external disturbances.

**Adaptability and Multiband Functionality:** Fractal antennas can be designed to cover multiple frequency bands simultaneously, allowing for seamless communication across different wireless technologies or standards. This adaptability is advantageous in remote patient monitoring, as it enables compatibility with various wireless protocols, such as Bluetooth, Wi-Fi, or cellular networks, without the need for separate antennas for each frequency band. Overall, fractal antennas facilitate compact design, wideband capability, enhanced signal strength and coverage, reduced interference, and adaptability in remote patient monitoring. These antennas enable seamless wireless communication, ensuring reliable transmission of vital signs data and facilitating effective remote monitoring of patient's health conditions.

## 5.9 ADVANCED SIGNAL PROCESSING TECHNIQUES FOR ENHANCED EFFICIENCY AND RELIABILITY

### 5.9.1 Signal processing techniques for enhancing data transmission

#### *5.9.1.1 Optimizing data transmission*

"Advanced Signal Processing Techniques for Enhanced Efficiency and Reliability" explores the pivotal role of signal processing in elevating the performance of data transmission systems. In the realm of communication technology, this approach involves the strategic manipulation and analysis of signals to improve their overall quality, mitigate interference, and counteract factors such as distortion and attenuation. The application of advanced modulation schemes, error correction codes, and equalization algorithms plays a vital role in enhancing the capabilities of data transmission. Furthermore, sophisticated filtering and noise reduction methods are employed to minimize undesirable signals and optimize the signal-to-noise ratio, ensuring the accurate and distortion-free reception of transmitted data. This comprehensive utilization of signal processing techniques results in heightened throughput, increased reliability, and improved overall performance in diverse communication environments, offering a substantial advancement in the efficiency of data transmission systems.

### 5.9.2 Noise reduction and signal integrity in medical applications

In medical applications, ensuring the accuracy and reliability of transmitted data is of utmost importance, and this is addressed through dedicated efforts in noise reduction and maintaining signal integrity. The sensitivity of medical equipment and the critical nature of the data involved necessitate advanced signal-processing techniques. Noise reduction plays a pivotal role in minimizing interference that may arise from various sources, such as electronic equipment or environmental factors, ensuring that the medical signals remain clear and undistorted. Concurrently, preserving signal integrity is crucial to prevent data corruption during transmission. Techniques like advanced filtering, shielding, and error correction are employed to enhance the robustness of the transmitted signals, safeguarding the accuracy of vital medical information. By focusing on noise reduction and signal integrity in medical applications, the medical community can significantly improve the precision of diagnostic tools, monitoring devices, and other critical systems, ultimately contributing to more reliable healthcare outcomes.

## 5.10 CHALLENGES AND LIMITATIONS

**Complex Design and Analysis:** The intricate geometry of fractal antennas, while offering unique advantages, can pose challenges in terms of design complexity and analytical modeling, making their implementation and optimization more challenging.

**Bandwidth Considerations:** While fractal antennas often exhibit broad bandwidth characteristics, there can be challenges in achieving optimal performance across all desired frequency ranges, potentially limiting their applicability in certain medical applications that require specific frequency bands.

**Size Constraints:** Fractal antennas are known for their compact size, but in certain medical devices or scenarios, strict size constraints may pose challenges in integrating these antennas without compromising their performance.

**Environmental Interference:** The effectiveness of fractal antennas can be influenced by the surrounding environment, and factors such as interference from nearby electronic devices or physical obstacles may affect their performance, especially in dynamic medical settings.

**Manufacturing Challenges:** Fabricating fractal antennas with precision can be challenging, and the manufacturing process may introduce variability that could impact their overall performance and reliability.

**Power Consumption:** In some medical applications where power efficiency is critical, the inherent complexity of fractal antennas may lead to increased power consumption, posing challenges for battery-operated devices.

**Limited Directionality:** Fractal antennas may exhibit less directional control compared to traditional antennas, limiting their suitability for certain medical applications that require precise signal focusing or beamforming.

While recognizing the remarkable potential of fractal antennas in medicine, addressing these challenges and limitations is essential for maximizing their benefits and ensuring their successful integration into medical devices and communication systems. Researchers and practitioners need to consider these factors to harness the advantages of fractal antennas effectively within the healthcare domain.

### 5.10.1 Regulatory challenges in the approval of medical devices with fractal antennas

Navigating the approval process for medical devices featuring fractal antennas is accompanied by a set of distinctive regulatory challenges. The integration of fractal antennas into medical devices introduces complexities that may not be explicitly addressed in existing regulatory frameworks. Regulatory bodies, such as health authorities and agencies overseeing medical device approvals, may face challenges in comprehensively evaluating the safety and efficacy of these devices due to the unique geometric characteristics and unconventional design principles of fractal antennas. Ensuring compliance with established standards and demonstrating the reliability of fractal antennas under diverse operating conditions becomes crucial. Additionally, the need for standardized testing methodologies specific to fractal antennas may arise, demanding collaboration between industry stakeholders and regulatory bodies to establish robust evaluation criteria. Addressing these regulatory challenges is imperative to facilitate the seamless integration of medical devices incorporating fractal antennas into the healthcare landscape, ensuring both innovation and adherence to safety standards in the approval processes.

### 5.10.2 Integration issues with existing medical technologies

The integration of medical devices featuring fractal antennas poses notable challenges when interfacing with existing medical technologies. The unconventional design and unique geometric properties of fractal antennas may require adaptations to ensure seamless compatibility with established medical equipment and systems. Standardized interfaces and communication protocols within the healthcare infrastructure may need to be modified or extended to accommodate the distinctive characteristics of fractal antennas. Additionally, interoperability concerns may arise, particularly when integrating these antennas with legacy devices that were not originally designed to interface with such advanced technologies [17]. Calibration and synchronization issues could emerge, impacting the accuracy and reliability of data transmission between devices. Collaborative efforts between manufacturers, healthcare providers, and regulatory bodies are essential to establish guidelines and standards that address these integration challenges. By addressing these issues head-on, the healthcare industry can effectively incorporate fractal antenna-equipped medical devices, fostering a cohesive and interconnected healthcare ecosystem.

## 5.11 FUTURE DIRECTIONS AND INNOVATIONS

### 5.11.1 Emerging trends and future possibilities in fractal antenna research for medicine

The realm of fractal antenna research for medical applications is witnessing the emergence of several promising trends and presents a plethora of future possibilities. One notable trend involves the continued exploration of novel fractal geometries and designs to enhance the performance of medical antennas. Researchers are actively investigating how intricate fractal patterns can be tailored to specific medical applications, optimizing characteristics such as bandwidth, efficiency, and size. Another significant trend is the integration of fractal antennas into miniaturized and wearable medical devices, paving the way for unobtrusive yet powerful technologies for patient monitoring, diagnostics, and treatment.

Moreover, the convergence of fractal antennas with other cutting-edge technologies such as the Internet of Things (IoT) and artificial intelligence is opening new avenues. Fractal antennas can contribute to the development of smart and connected healthcare systems, facilitating real-time data transmission, remote patient monitoring, and personalized medicine. The utilization of fractal antennas in wireless biomedical sensing and imaging is gaining momentum, promising breakthroughs in non-invasive diagnostics and imaging technologies.

Looking ahead, the possibilities in fractal antenna research for medicine extend to areas such as wireless brain-machine interfaces, implantable medical devices, and advanced telemedicine applications. The exploration of biocompatible materials for fractal antennas may also lead to innovations in implantable devices that seamlessly integrate with the human body. Additionally, advancements in energy harvesting using fractal antennas may contribute to the development of self-powered medical implants, reducing the dependence on external power sources.

As research in fractal antennas for medical applications continues to evolve, collaboration between engineers, medical professionals, and regulatory bodies will be essential to ensure the successful translation of these innovations from the laboratory to clinical practice. The future holds exciting possibilities for the role of fractal antennas in revolutionizing medical technology, offering solutions that are not only technologically advanced but also tailored to the unique challenges and opportunities in the field of healthcare.

### 5.11.2 Potential advancements and interdisciplinary collaborations

The exploration of potential advancements in fractal antenna applications for medicine holds great promise, particularly through interdisciplinary collaborations that can harness diverse expertise. One avenue for advancement lies in the integration of fractal antennas with emerging technologies like 5G and beyond, enabling high-speed, low-latency communication for real-time medical data transfer. Collaborations between antenna engineers, telecommunications experts, and healthcare professionals can drive the development of seamless and reliable communication networks within medical environments.

Interdisciplinary collaborations also offer opportunities for advancing imaging technologies. By combining fractal antennas with advancements in medical imaging, such as magnetic resonance imaging (MRI) and computed tomography (CT), researchers can potentially enhance the resolution, sensitivity, and speed of imaging systems. Radiologists, antenna engineers, and materials scientists working together can contribute to the development of more efficient and accurate diagnostic tools.

Another exciting prospect involves the collaboration between researchers in material science and fractal antenna design to explore biocompatible materials. Developing antennas [18] that are compatible with the human body can lead to innovations in implantable medical devices and wearable technologies. By bringing together experts in biotechnology, materials science, and medical device design, interdisciplinary teams can drive progress in creating devices that seamlessly integrate with the human body.

Furthermore, advancements in energy harvesting technologies, coupled with interdisciplinary efforts, could lead to the development of self-powered medical devices. Collaboration between experts in energy harvesting, electronics, and medical applications can pave the way for implantable devices that sustain themselves through energy derived from the surrounding environment or the human body.

In summary, potential advancements in fractal antenna applications for medicine are closely tied to interdisciplinary collaborations. These collaborations can propel innovations in communication networks, imaging technologies, biocompatible materials, and energy harvesting, contributing to the development of more efficient, reliable, and patient-friendly medical devices and systems. As researchers from diverse fields work together, the future holds the promise of transformative breakthroughs at the intersection of fractal antennas and medical technology.

## 5.12 CASE STUDIES

### 5.12.1 Real-world examples of successful implementation of fractal antennas in medical settings

**Introduction:** The implementation of fractal antennas in medical settings has brought about transformative advancements in connectivity, diagnostics, and patient care. This case study explores real-world examples of successful applications of fractal antennas, showcasing their impact on medical technology.

**Case 1: Wireless Biomedical Sensing System for Continuous Monitoring:**

*Objective:* To demonstrate the integration of fractal antennas in a wireless biomedical sensing system for continuous patient monitoring.

*Scenario:* In a hospital setting, a wireless biomedical sensing system is deployed to monitor the vital signs of patients in real-time. Fractal antennas are incorporated into wearable sensors, ensuring reliable and high-speed communication with centralized monitoring stations. The fractal antenna's compact design and enhanced signal processing capabilities enable seamless data transmission, providing healthcare professionals with accurate and timely information for critical decision-making.

**Case 2: Improved Imaging Resolution in MRI Diagnosis:**

*Objective:* To illustrate the impact of fractal antennas on enhancing imaging resolution in magnetic resonance imaging (MRI).

*Scenario:* A diagnostic imaging center adopts fractal antennas in its MRI machines to improve signal strength and resolution. The unique geometric properties of fractal antennas contribute to clearer imaging, enabling more precise diagnoses. The implementation results in enhanced imaging quality, reducing the need for repeated scans and improving overall efficiency in the diagnostic process.

**Case 3: Advancements in Telemedicine Connectivity:**

*Objective:* Showcase the use of fractal antennas in telemedicine applications for remote patient care.

*Scenario:* A telemedicine initiative employs fractal antennas to establish robust and high-bandwidth connections between patients in remote locations and healthcare professionals. This implementation facilitates seamless video consultations, real-time monitoring, and secure data transmission, bridging geographical gaps and expanding access to quality healthcare services.

**Case 4: Enhanced Communication in Implantable Medical Devices:**

*Objective:* Demonstrate the successful integration of fractal antennas in implantable medical devices for efficient communication within the body.

*Scenario:* A medical device company incorporates fractal antennas into implantable devices such as neuro stimulators. The fractal antennas enable reliable wireless communication, optimizing the control and programming of the devices. This implementation improves patient outcomes by allowing healthcare providers.

### 5.12.2 Lessons learned

**Interference Management:** Anticipating and addressing interference sources, particularly in hospital environments, is crucial for the reliable operation of fractal antenna-enabled systems.

**Adaptation to Existing Protocols:** Integration of fractal antennas into established medical protocols requires thoughtful consideration and potential adjustments to ensure seamless compatibility.

**Customization for Diverse Environments:** The adaptability of fractal antenna designs is essential, especially in diverse healthcare settings, such as remote areas with limited infrastructure.

**Energy Efficiency in Implantable Devices:** Careful management of power consumption is vital in the implementation of fractal antennas in implantable medical devices to optimize battery life and overall device functionality.

**Conclusion:** The successful implementation of fractal antennas in medical settings has provided valuable insights. Lessons learned from addressing interference, adapting to existing protocols, customizing for diverse environments, and optimizing energy efficiency contribute to the ongoing refinement and advancement of fractal antenna applications in healthcare. These lessons serve as a foundation for future innovations, ensuring that fractal antennas continue to play a pivotal role in transformative connectivity within the medical landscape.

## 5.13 CONCLUSION AND RECOMMENDATIONS

### 5.13.1 Summarizing key findings from the comprehensive review

The integration of fractal antennas in medicine represents a transformative advancement in communication and connectivity within healthcare applications. The comprehensive analysis of design, performance, advantages, and disadvantages underscores the unique characteristics and potential of fractal antennas to enhance medical technology. The multiband capabilities, compact design, wide bandwidth, and customization options position fractal antennas as valuable components in various medical devices, ranging from implantable devices to wireless biomedical sensing systems.

The advantages of fractal antennas, such as their ability to improve signal-to-noise ratio, adaptability to dynamic frequency changes, and inherent noise reduction capabilities, make them particularly well-suited for the evolving landscape of medical technology. These features address key challenges in medical settings, contributing to more efficient and reliable communication, diagnostic imaging, and patient monitoring. However, the complexity of fractal antenna design, potential integration challenges with existing technologies, and considerations related to power consumption and environmental sensitivity must be carefully navigated.

## REFERENCES


1. Chowdary, P. S. R., A. M. Prasad, P. M. Rao, and J. Anguera. "Design and performance study of sierpinski fractal based patch antennas for multiband and miniaturization characteristics." *Wireless Personal Communications* 83, no. 3 (2015): 1713–1730.
2. Romeu, J., and J. Soler. "Generalized Sierpinski fractal multiband antenna." *IEEE Transactions on Antennas and Propagation* 49, no. 8 (2001): 1237–1239.
3. Abdelati, R. E. H. A., E. L. Abdelkebir, O. Benhmammouch, and A. OULAD SAID. "Fractal antennas: A novel miniaturization technique for wireless networks." *Transactions on Networks and Communications* 2, no. 5 (2014): 165–193.
4. Werner, D. H., and S. Ganguly. "An overview of fractal antenna engineering research." *IEEE Antennas and Propagation Magazine* 45, no. 1 (2003): 38–57.
5. Balanis, C. "Chapter 4: Linear wire antennas". In *Antenna Theory Analysis and Design,* Third ed., Hoboken, NJ: John Wiley & Sons, Inc, pp. 151–219 (2005).
6. (a) Azari, A. "A new fractal antenna for super wideband applications." In *PIERS Proceedings*, pp. 885–888, 2010.; (b) Krzysztofik, W. J. "Take advantage of fractal geometry in the antenna technology of modern communications." In *2013 11th International Conference on Telecommunications in Modern Satellite, Cable and Broadcasting Services (TELSIKS)*, Nis, Serbia, 2013.
7. Lizzi, L., R. Azaro, G. Oliveri, and A. Massa, "Multiband fractal antenna for wireless communication systems for emergency management". *Journal of Electromagnetic Waves and Applications*, 26(1), pp. 1–11, 2012.
8. Dinesh, V. and G. Karunakar, "Analysis of microstrip rectangular carpet shaped fractal antenna". In *Signal Processing and Communication Engineering Systems (SPACES), 2015 International Conference on*, IEEE, Vijayawada, pp. 531–535, 2015.
9. Patil, S., and V. Rohokale. "Multiband smart fractal antenna design for converged 5G wireless networks." In *Pervasive Computing (ICPC), 2015 International Conference on*, IEEE, Pune, pp. 1–5, 2015.
10. Raval, B. T., P. R. Pimpalgaonkar, M. R. Chaurasia, and T. Upadhyaya. Review of ultra-wideband and design studies of patch antenna for ultra-wideband communication, ICAI 2016, Kota, 2016.
11. Ahmed A. K., Al-Zabee, S. Qasim Jabbar, and D. Wang. "Fractal antennas (study and review)". *International Journal of Computers and Technology*, 15(13), 256–265 (2018). ISSN 2277-3061.

12. Werner, D. H., and S. Ganguly. "An overview of fractal antenna engineering research." *IEEE Antennas and propagation Magazine*, 45(1), pp. 38–57, 2003.
13. Patil, S., and V. Rohokale. "Multiband smart fractal antenna design for converged 5G wireless networks." In *2015 International Conference on Pervasive Computing (ICPC)*, pp. 1–5. IEEE, India, 2015.
14. Takebe, K., H. Miyashita, K. Takano, M. Hangyo, and S.-S. Lee. "Electromagnetic wave absorption characteristics of H-shaped fractal antenna for multi-band microbolometer." In *2014 9th IEEE International Conference on Nano/Micro Engineered and Molecular Systems (NEMS)*, 13–16 April 2014, Waikiki Beach, Hawaii. Doi:10.1109/NEMS.2014.6908776
15. A. Kaur and G. Singh, "A review paper on fractal antenna". *International Journal of Advanced Research in Electrical, Electronics and Instrumentation Engineering, (An ISO 3297: 2007 Certified Organization)*, 3(6), 265–272, 2014.
16. M. Levy, S. Bose, A. Dinh, and D. S. Kumar, "A novelistic fractal antenna for ultra wideband (UWB) Applications", *Progress in Electromagnetics Research B*, 45, 369–393, 2012.
17. M. Levy, D. S. Kumar, and A. V. Dinh, "Analysis of nonlinear fractal optical antenna arrays: A conceptual approach", *Progress in Electromagnetics Research B*, 56, 289–308, 2013.
18. M. Levy, D. S. Kumar, and A. Dinh, "Analysis of nonlinear fractal optical antenna arrays: A conceptual approach", *Progress in Electromagnetics Research M*, 32, 83–93, 2013.

Chapter 6

# Design of flexible antenna using felt substrate for ISM band applications

*M. Pandimadevi, R. Tamilselvi, and M. Parisa Beham*

## 6.1 INTRODUCTION

In recent years, there has been a great deal of interest in the field of flexible electronic systems from both academic and industrial sources [1]. Flexible technology is desirable for the future electronics world because of its lower weight, cheaper manufacture, easier fabrication, and the availability of low-cost stretchable materials such as cotton, felt, polyester, foams, denim, and so on. Furthermore, recent advances in the production of tiny, flexible wireless parts have prepared the way for the immortalization of such adaptable systems. In today's environment, there are a variety of antenna types. Yagi Uda, loop, parabolic, and microstrip patch antennas are examples. Each type has its own set of benefits and drawbacks. Among these, the patch antenna is the most widely used radiating element in stretchy antenna designs. The patch antenna has risen to the front of the pack among all antennas due to its compact size and simple design. Patch antennas [2] have a few advantages over traditional antennas, which are as follows:

- Lightweight and low-profile, with the ability to conform to curved surfaces.
- Cheaper.
- MICs (microwave integrated circuits) are easily integrated.
- Easy to fabricate.

Recently, antennas [3] have been made using many textile and synthetic substrate materials such as cotton, polyester, jeans, cellulose (paper-based), pulp fibres, rubber, foam, flannel fabric, etc. These materials make the antenna more adjustable in all environments. Of the two (textile and synthetic material), textile materials are more often readily available and easy to deploy.

The combination of communication devices into textiles is rapidly increasing in commercial as well as military applications [4]. Rapid advancements in computer networks, such as personal area network, body area network,

 DOI: 10.1201/9781003560487-6

and local area networks, have enabled more expeditious and more precise data transfer and analysis. Technologies such as Radio and TV broadcasting, cellular communication (Bluetooth, 3G/4G), Global Positioning System, Wi-Fi, WiMAX, Radio Frequency Identification, and Radar are being integrated into a single system called communication and network systems. Many researches were carried out on flexible textile antennas. The following are some of the works focussed on flexible antennas using textile material as substrates.

Indumathi and Bhavithra [5] designed an antenna with Jeans fabric as the substrate for patient monitoring applications. Four types of antennas were designed, fabricated, and compared by the authors. There is a reduction in the return loss of –29 dB with higher substrate dimensions.

Asim Ali Khan et al. [6] show the design of an antenna using pulp fibres (Lignocellulose fibril sheet) as a substrate for dual-band operation at 2.4–2.5 and 5.725–5.875 GHz. The pulp-based antenna operates with measured return loss values of –15 dB at 2.7 GHz and –28 dB at 5.6 GHz. Though the antenna shows better results with respect to $S_{11}$ (–28 dB), there is no proof about its bending and wet performance, and the substrate dimensions are higher.

Saadat Hanif Dar et al. [7] designed a flexible wearable antenna using a rubber substrate. In this work, bandwidth decreases with the addition of content in rubber material.

Reema Dubey et al. [8], designed trophy-shaped flexible wearable antenna using foam substrate material whose dielectric constant is approximately equal to unity. Three different shapes of antenna using foam substrate were designed, simulated and compared. The return loss was observed to be around –30 dB. However, there is a major shift in the operating frequency between the measured and simulated antenna.

Thus, all flexible substrate materials have their own characteristics such as disability to absorb moisture, high cost, unavailability, and bending effects that affect the overall performance of the antenna. Recently, it has been exhibited that Felt material can act as substrate material for the microstrip antennas. Felt [9] is a type of textile made by matting, condensing, and pressing threads together. Felt can be made from natural fibres like wool or animal fur, or synthetic fibres like rayon made from wood pulp. It's fire-resistant and self-extinguishing, dampens vibrations and absorbs sound, and can hold a lot of liquid without getting wet. Felt material has dielectric constant $\varepsilon_r = 1.45$ and Loss Tangent=0.02. This work is the initial attempt to display the utilization of Felt as a suitable substrate for flexible antennas at ISM band frequencies.

The rest of the chapter has been organized as follows: The antenna design section deals with the design of the proposed antenna followed by antenna simulation, a discussion of results and then a conclusion.

## 6.2 ANTENNA DESIGN

For designing the microstrip patch antenna [10], the resonant frequency, thickness, and dielectric constant values of the substrate material should be chosen, and the width of the patch *W* is calculated as 50 mm from Eq. (6.1) as,

$$W = \frac{1}{2f_r\sqrt{\mu_0\varepsilon_0}}\sqrt{\frac{2}{\varepsilon_r + 1}} \tag{6.1}$$

The effective dielectric constant ($\varepsilon_{reff}$) is defined as a function of the ratio of the width to the height of a microstrip line (W/h), as well as the dielectric constant of the substrate material. It can be calculated as

$$\varepsilon_{reff} = \frac{\varepsilon_r + 1}{2} + \frac{\varepsilon_r - 1}{2}\left[1 + 12\frac{h}{W}\right]^{-1/2} \tag{6.2}$$

Where *h* is the height of the dielectric substrate material,

The effective dielectric constant was calculated as 1.05 by substituting the height of the substrate as 1 mm, the width of the patch as 50 mm, and the relative permittivity of the substrate as 1.45.

The effective length of the patch ($L_{eff}$) is the sum of the actual length of the antenna and its extension or the fringe effects. It is calculated as

$$L_{eff} = \frac{c}{2f_r\sqrt{\varepsilon_{reff}}} \tag{6.3}$$

The effective length of the patch was calculated as 42.76 mm.

The length extension can be calculated from

$$\Delta L = 0.412h\frac{(\varepsilon_{reff} + 0.3)\left(\frac{W}{h} + 0.264\right)}{(\varepsilon_{reff} - 0.258)\left(\frac{W}{h} + 0.8\right)} \tag{6.4}$$

The length extension was calculated as 1.815 mm.

The actual length of the patch is obtained by using Eq. (6.5) as

$$L = L_{eff} - 2\Delta L \tag{6.5}$$

Finally, the length of the patch was calculated as 45 mm.

Figure 6.1 represents the structure of the proposed antenna. Using the equation, the parameters are calculated. The calculated parameters are listed below in Table 6.1. The copper is used for the patch and strip line. As the patch is square with a slot shape, the side length is 45 mm, and the thickness is 1.1 mm.

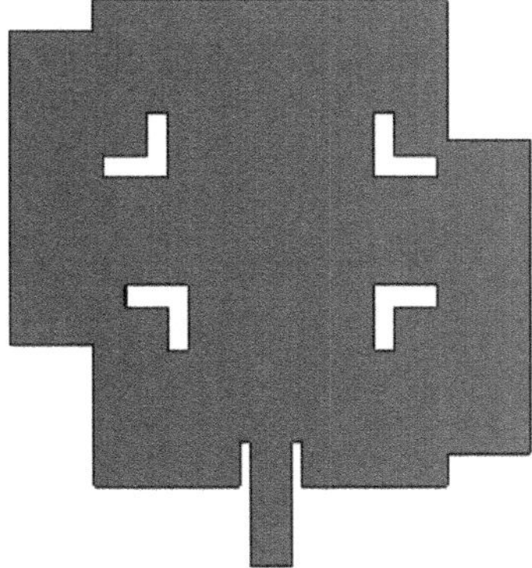

*Figure 6.1* Structure of proposed antenna.

*Table 6.1* Proposed antenna dimension

| *Parameters* | *Dimension (mm)* |
|---|---|
| Patch length and width | 45×50 |
| Substrate | 60×60 |
| Patch thickness | 0.1 |
| Ground plane thickness | 0.1 |
| Substrate thickness | 1 |

This antenna is made out of a low-dielectric-constant substrate felt material. The square with slot and *L* slots make up the antenna was excited from port 1 via a 50 Ω coaxial probe transmission line. The flexible microstrip patch antenna is made from "Felt," which is a flexible substrate. Felt is a material with a low dielectric constant, resulting in minimal return loss. The ground plane, patch, and strip line are all made of copper. The foam sheet is fixed as the substrate on the ground plane, with copper as the ground plane. An antenna's slot is used to reduce the antenna's size. An antenna's gain is improved when its size is reduced. Copper is used to make the patch and strip lines, which are then adhered to the foam substrate. The antenna is meant to operate at a frequency of 2.45 GHz.

## 6.3 ANTENNA SIMULATION

The design was simulated in Computer Simulation Technology (CST) Studio suite software [11] with varying the substrate material. Figure 6.2 shows the proposed design using a Jute substrate in CST software. The ISM band operating frequency is achieved with a return loss of –25.63, as shown in Figure 6.3. The VSWR, Gain, Directivity, and Radiation Pattern measurements are shown in Figures 6.4–6.7, respectively.

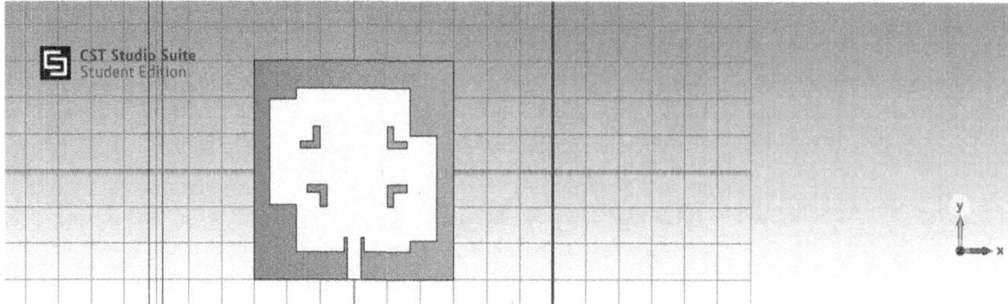


*Figure 6.2* Design of antenna using felt substrate.

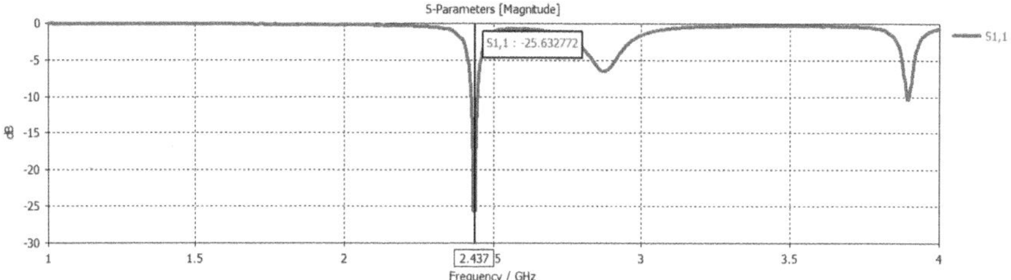


*Figure 6.3* Returns loss measurement.

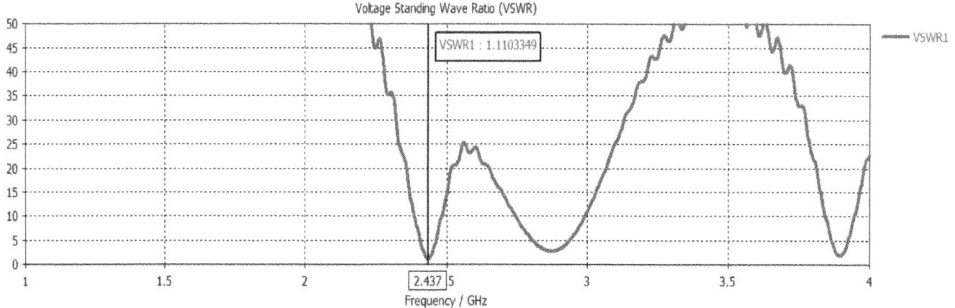


*Figure 6.4* VSWR measurement.

## 6.4 RESULT AND DISCUSSIONS

Table 6.2 shows a comparison of simulation results of the antenna using various substrate materials. In this, when compared to all other materials Felt shows better results in terms of Gain (6.7 dBi) and return loss (–25.63 dB). While antennas using Polyester [12] and Pulp [6] materials also exhibit good gain values (7.81 and 6.94 dBi), their larger dimensions and higher return losses can degrade overall antenna performance.

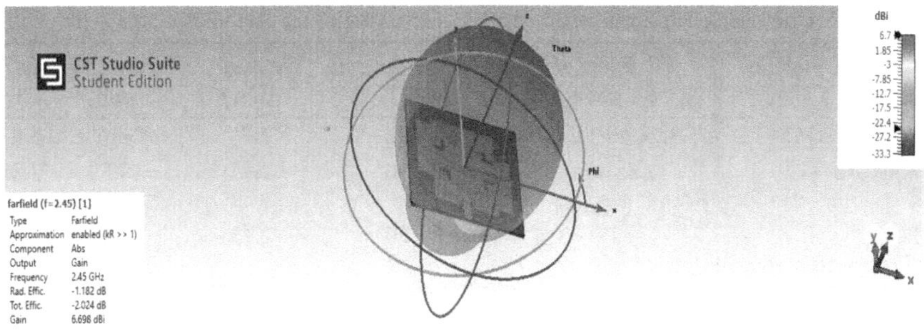


*Figure 6.5* Gain measurement.

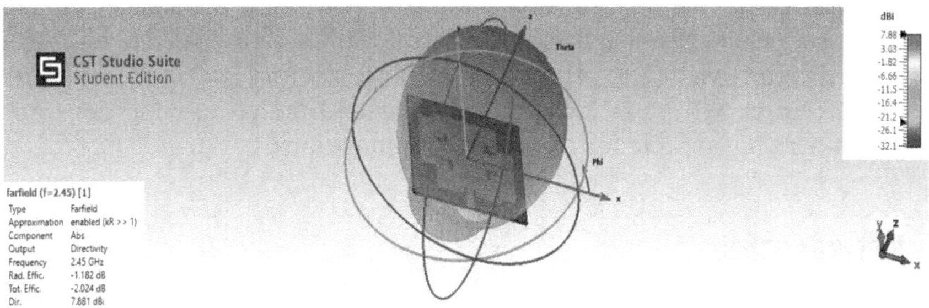


*Figure 6.6* Directivity measurement.

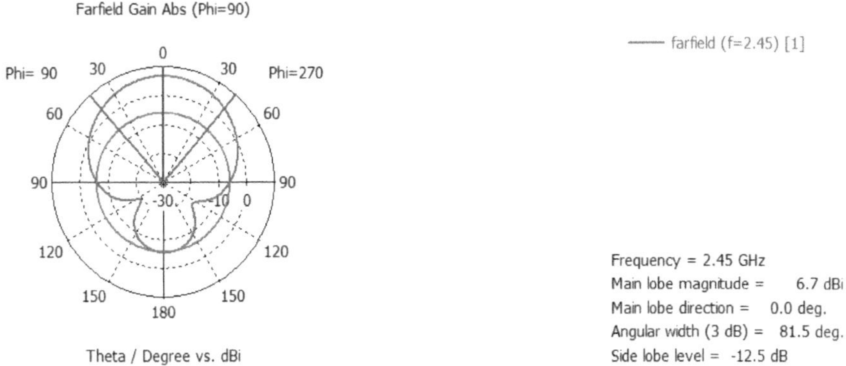


*Figure 6.7* Radiation pattern.

*Table 6.2* Comparison of simulation results of antenna using various substrate material

| *Substrate material used* | *Thickness of the substrate* | *Size of the antenna ($mm^2$)* | *Frequency (GHz)* | *Return loss $S_{11}$(dB)* | *Gain (dBi)* |
|---|---|---|---|---|---|
| Pulp fibre [6] | Around 1 mm | 58 × 82 | 2.7 | <−15 | 6.94 |
| Polyester [12] | 2.85 mm | 90 × 90 | 2.45 | −10.52 | 7.81 |
| Felt (Proposed material) | 1 | 60 × 60 | 2.45 | −25.63 | 6.7 |

## 6.5 CONCLUSION

The proposed antenna presented is very versatile and it is easy to make it operate at ISM frequency bands. In addition, an antenna using felt substrate will improve the bandwidth thus suitable for wearable antennas too. The simulation results (Return loss (–25.63 dB), VSWR (less than 2), Gain (6.7 dBi) and Directivity (7.88 dBi)) were also satisfactory. In future, the proposed antenna will be fabricated and measured under all conditions and will make them suitable for commercial applications.

## REFERENCES

1. H. R. Khaleel, H. M. Al-Rizzo and A. I. Abbosh, Design, fabrication, and testing of flexible antennas, *Advancement in Microstrip Antennas with Recent Applications*, 2013, doi: 10.5772/50841.
2. M. Pandimadevi and R. Tamilselvi, Design issues of flexible antenna: A review, *International Journal of Advanced Trends in Computer Science and Engineering*, pp. 1386–1394, 2019, doi: 10.30534/ijatcse/2019/55842019.
3. H. Yongan, J. Chen, Z. Yin, and Y. Xiong, Roll-to-roll processing of flexible heterogeneous electronics with low interfacial residual stress, components, packaging and manufacturing technology, *IEEE Transactions*, Vol. 1, No. 9, pp. 1368–1377, 2011.
4. I. Singh and V. S. Tripathi, Micro strip patch antenna and its applications: A survey, *International Journal of Computer Applications in Technology*, Vol. 2, No. 5, pp. 1595–1599, 2011.
5. G. Indumathi and J. Bhavithra, Design of wearable textile antenna for patient monitoring, *International Journal of Pure and Applied Mathematics*, Vol. 119, No. 7, pp. 1179–1184, 2018.
6. A. A. Khan, A. Razaq, J. Ali, F. Arshad, M. Mumtaz, and S. Khan, Antenna miniaturization using pulp fibers as a substrate for dual band operation, *Microwave and Optical Technology Letters*, Vol. 58, No. 9, pp. 2146–2148, 2016, doi: 10.1002/mop.30002.

7. S. H. Dar, J. Ahmed, and M. Raees, Characterizations of flexible wearable antenna based on rubber substrate, *International Journal of Advanced Computer Science and Applications (IJACSA)*, Vol. 7, No. 11, 2016, doi: 10.14569/IJACSA.2016.071124.
8. R. Dubey, V. K. Singh, A. Kumar, and B. Z. Ali, Realization of trophy based flexible wearable antenna based on foam substrate, *International Journal of Engineering & Technology*, Vol. 7, No. 2.12, pp. 222–224, 2018.
9. https://wunderlabel.com/blog/p/sewing-felt-pros-cons/.
10. M. Pandimadevi and R. Tamilselvi, Comparative study of micro strip patch antenna using different flexible substrate materials, *International Journal of Pure and Applied Mathematics*, Vol. 118, No. 22, pp. 751–766, 2018.
11. https://www.3ds.com/products-services/simulia/products/cst-studio-suite/.
12. S. Hussain, S. Hafeez, S. A. Memon, and N. Pirzada, Design of wearable patch antenna for wireless body area networks, *International Journal of Advanced Computer Science and Applications (IJACSA)*, Vol. 9, No. 9, pp. 146–151, 2018.

Chapter 7

# Design and analysis of foam-based flexible wearable MSA for healthcare applications

*Kailash V. Karad, Vaibhav S. Hendre, Jaswantsing L. Rajput, and Mandar P. Joshi*

## 7.1 INTRODUCTION: BACKGROUND

Wearable antennas have generated a lot of interest recently because of their eye-catching features and potential to enable flexible, lightweight, affordable, and portable wireless communication and sensing. When utilized on various parts of the human body, such antennas must be conformal, which necessitates the use of flexible materials and a low-profile structure. Eventually, these antennas must be able to operate near the human body for the least amount of deprivation. Wearable antennas are difficult to design because of these requirements, especially when considering factors like size, compactness, coupling to the body, structural deformation, fabrication complexity, and accuracy. In spite of minor differences in applications, the majority of these issues occur in the context of body-worn implementation. Nowadays, microstrip antennas (MSA) are widely used for reasons like low profile, lightweight, versatility, and companionable with integrated circuits [1]. In several applications, wireless local area networks (WLAN), wireless interoperable microwave access (WiMAX), long-term evolution (LTE) [2,3], and mmWave [4] MSA are necessary.

In general, wearable antennas are made expressly to operate on the body when worn. This necessitates the use of small, flexible antennas. Wearable antennas have been employed in a wide variety of applications including object observation and wireless medical applications because of their flexibility and robustness. The ISM bands need little power and operate without a license. As a result, numerous studies have been conducted to integrate ISM bands into a single device [5]. Traditional antennas are not suitable for wearable applications since they are renowned for being hard and inflexible [6]. On the other hand, wearable antennas must be flexible to bend and follow the contours of the user's body. They also need to be compact and lightweight because they are intended to be sewn into garments [7]. For flexible and wearable applications, polymer-based dielectric materials have also recently attracted considerable interest. Polytetrafluoro Ethylene (PTFE), Kapton polyimide, Polyethylene dioxythiophene (PEDOT), and polydimethylsiloxane (PDMS) substrates are a few examples


DOI: 10.1201/9781003560487-7

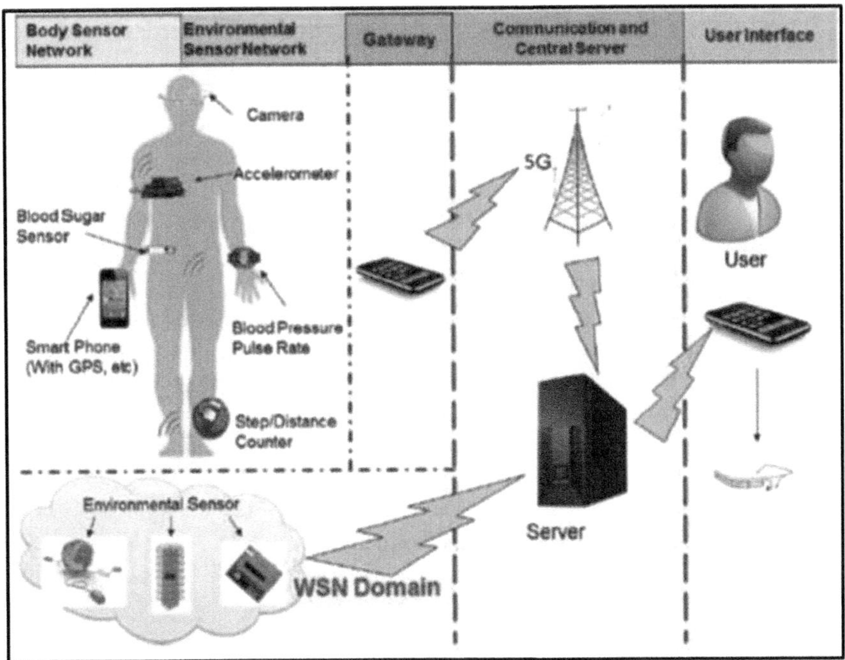


*Figure 7.1* Wearable antenna applications with sensors mounted on the body [13].

of polymer-based materials that are used nowadays for flexible wearable antennas [8–11]. They are a suitable choice for flexible body-worn applications because of their enhanced flexibility, reduced humidity absorption, and small loss tangent [12]. Figure 7.1 shows the applications of wearable antennas with sensors mounted on the body and different communication media used for signal transmission.

However, a requirement to use inexpensive, broadly accessible substrates in the creation of innovative structures for wearable applications is reinforced by the relatively greater cost and limited commercial availability of specialized substrates. A flexible nature of the foam material is employed in this study. Thus, a flexible wearable antenna for healthcare applications is presented as a remedy to the rigidity and inflexibility of conventional antennas.

A wearable antenna is a specialized antenna designed to be integrated into clothing, accessories, or other items that can be worn on the body. Here's an overview of wearable antennas at a glance [4,8,10,11,14–17].

1. **Purpose:** Wearable antennas are used to facilitate wireless communication between devices, such as smartphones, smartwatches, fitness trackers, and other wearable technology, without the need for external protruding antennas.

2. **Form Factor:** They come in various form factors, including:
    - **Fabric-Based Antennas:** Made from conductive textiles or threads, these antennas can be seamlessly integrated into clothing, hats, gloves, or other wearable items.
    - **Flexible PCB Antennas:** These antennas are printed on flexible substrates, allowing them to conform to curved surfaces or bend with the body's movements.
    - **Miniature Antennas:** Designed to be compact and lightweight, these antennas are often embedded in small wearable devices or accessories.
3. **Integration:** Wearable antennas are typically integrated into the fabric or structure of the wearable item to minimize visibility and maximize user comfort. They may be sewn, adhered, or otherwise attached to the material.
4. **Frequency Bands:** Wearable antennas can operate across various frequency bands, including:
    - **Bluetooth:** for short-range communication between devices.
    - **Wi-Fi:** for wireless internet connectivity.
    - **Cellular:** for mobile communication (e.g., 2G, 3G, 4G, 5G).
    - **GPS:** for location tracking and navigation.
    - mmWave (28, 38, 60 GHz)
5. **Challenges**: Designing effective wearable antennas presents several challenges, including:
    - **Size Constraints:** Antennas must be small and lightweight to avoid discomfort and interference with movement.
    - **Performance:** Antennas must maintain sufficient radiation efficiency and bandwidth despite their compact size and integration into clothing.
    - **Flexibility:** Antennas should be flexible enough to conform to the shape of the body or wearable item without compromising performance.
    - **Environmental Factors:** Antennas must withstand exposure to sweat, moisture, and other environmental conditions encountered during daily wear.
6. **Applications:** Wearable antennas find applications in various fields, including:
    - **Fitness and Sports Monitoring:** for tracking biometric data such as heart rate, steps taken, and calories burned.
    - **Healthcare:** for remote patient monitoring and medical diagnostics.
    - **Industrial:** for asset tracking, personnel monitoring, and communication in hazardous environments.
    - **Military and First Responders:** for situational awareness, communication, and location tracking in the field.

## 7.2 ANTENNA DESIGN

The basic geometry of the microstrip patch antenna is rectangular, as depicted in Figure 7.2. A foam with a 3 mm thickness, a loss tangent of 0.0025, and a dielectric constant of 1.07 is used as a flexible substrate for the simulation of the antenna into the Ansys HFSS simulator. The foam is used as the substrate material because it is inexpensive, widely available, and flexible enough to blend in with the wearer's clothing. An antenna's radiating patch and ground plane are created using 0.05 mm thick copper foil. Ansys HFSS simulator simulates and optimizes the theoretically constructed antenna using Eqs. (7.1)–(7.7) [18] to obtain the desired results at an operating frequency of 5.8 GHz.

1. **Patch Width:**

$$W = \frac{c}{2f}\sqrt{\frac{2}{\varepsilon_r + 1}} \tag{7.1}$$

2. **Patch Length:**

$$L = L_{\text{eff}} - 2\Delta L \tag{7.2}$$

Where,

$$L_{\text{eff}} = \frac{c}{2f\sqrt{\varepsilon_{\text{reff}}}} \tag{7.3}$$

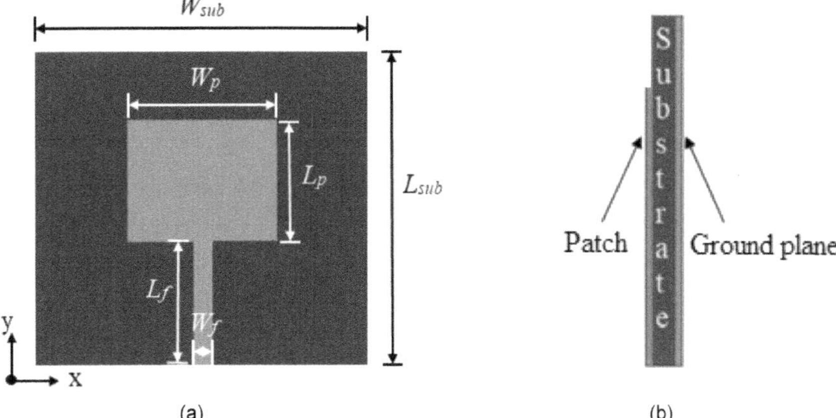


*Figure 7.2* Schematic of the proposed antenna.

and length extension is given by,

$$\Delta L = 0.412h \frac{(\varepsilon_{reff} + 0.3)\left(\frac{w}{h} + 0.264\right)}{(\varepsilon_{reff} - 0.258)\left(\frac{w}{h} + 0.8\right)} \tag{7.4}$$

$\varepsilon_{reff}$ = effective value of dielectric constant, and given as below,

$$\varepsilon_{reff} = \frac{\varepsilon_r + 1}{2} + \frac{\varepsilon_r - 1}{2}\left[1 + 12\frac{h}{w}\right]^{\frac{-1}{2}} \tag{7.5}$$

The patch impedance at the edge is approximated as,

$$Z_a \approx 90 \times \frac{\varepsilon_r^{\,2}}{(\varepsilon_r - 1)} \times \left(\frac{L}{W}\right)^2 \tag{7.6}$$

The impedance of transition width is given by,

$$Z_t = \sqrt{Z_0 \times Z_a} \tag{7.7}$$

The antenna design parameters are depicted in Table 7.1. The antenna is designed and simulated for the values considered.

The simulated antenna performance has been studied for return loss characteristic, VSWR and radiation pattern in terms of *E* and *H* plane and presented in Figure 7.3a–c. Whereas, an excellent impedance matching is observed as depicted in Figure 7.3d for the value of current distribution.

The antenna is resonating at 5.8 GHz and the value of VSWR is also found below 2 for the bandwidth of 470 MHz. A unidirectional radiation pattern is obtained in the *E* and *H* planes.

*Table 7.1* Design parameters

| *Parameter* | *Symbol* | *Value (mm)* |
|---|---|---|
| Patch length | $L_p$ | 19 |
| Patch width | $W_p$ | 24 |
| Substrate thickness | $H$ | 3 |
| Trace | $T$ | 0.050 |
| Feed length | $L_f$ | 20 |
| Feed width | $W_f$ | 3 |
| Ground length | $L_g$ | 50 |
| Ground width | $W_g$ | 54 |
| Substrate length | $L_s$ | 50 |
| Substrate width | $L_w$ | 54 |

## 7.3 PARAMETRIC ANALYSIS

The performance of the proposed antenna is analysed by varying the feed width '$W_f$' by 2 and 1 mm. The comparative parametric improvement is depicted in Table 7.2 and presented in Figure 7.4.

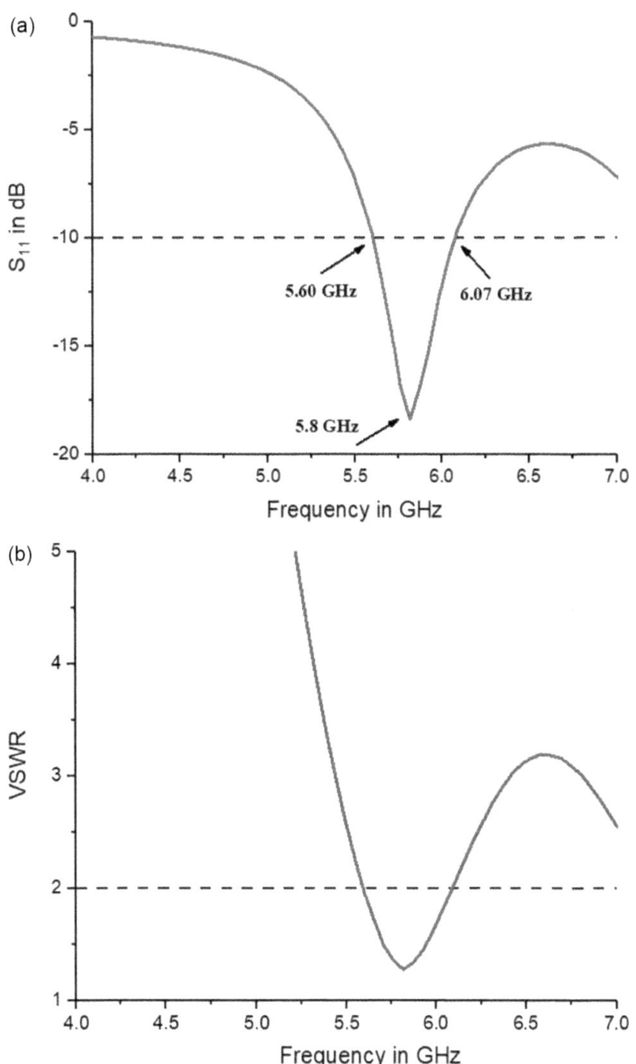


*Figure 7.3* Simulated results at 5.8 GHz.

*(Continued)*

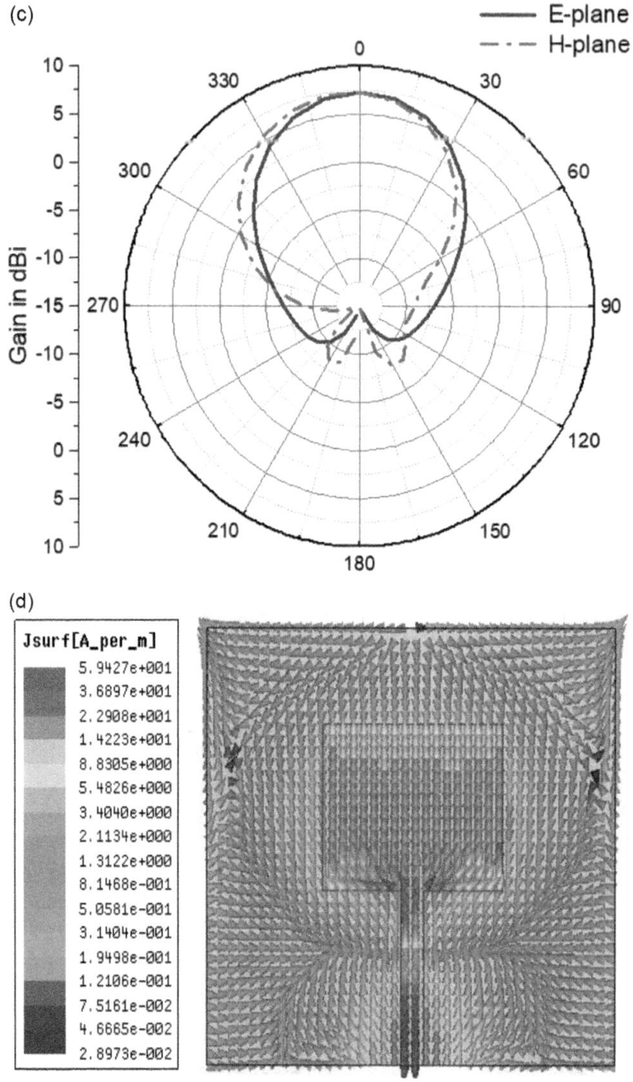

(c)

(d)

*Figure 7.3 (Continued)* Simulated results at 5.8 GHz.

*Table 7.2* Parametric variation for feed width of the proposed antenna

| *Feed width ($W_f$)* | *Return loss $S_{11}$ (dB)* | *Resonating frequency (GHz)* | *BW (MHz)* | *Gain (dBi)* |
|---|---|---|---|---|
| 3 mm | −18.75 | 5.8 | 470 | 7.2 |
| 2 mm | −26.78 | 5.9 | 463.4 | 7.4 |
| 1 mm | −15.51 | 5.9 | 327.9 | 8.0 |

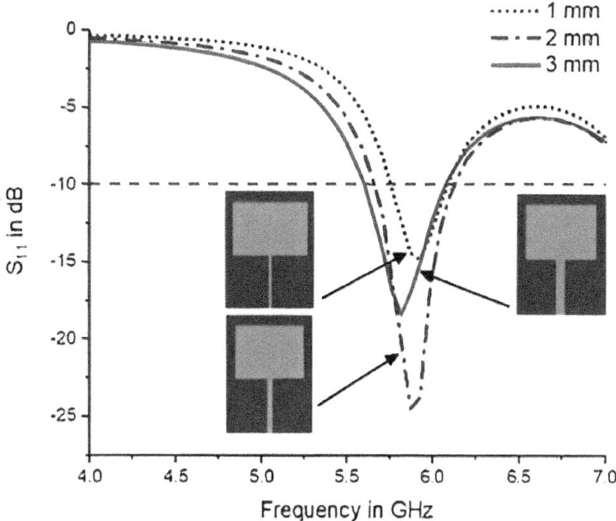


*Figure 7.4* Return loss ($S_{11}$) characteristic with variation in feed width $W_f$.

The results are presented in Table 7.2 shows the improvement in the value of bandwidth from 1 to 3 mm, whereas the gain value is decreased from 8 to 7.2 dBi respectively.

## 7.4 ON-BODY ANALYSIS

The on-body study offers valuable insights into the functionality of wearable technology, user engagement, and its effects on human physiology and behaviour, all of which contribute to the advancement of the wearable antenna. The on-body analysis performance of the proposed antenna is analysed to investigate the value of Specific Absorption Rate (SAR) for 1 and 10 g tissue of the human body [19–20]. SAR is defined as the power absorbed per unit mass of tissue, usually measured in watts per kilogram (W/kg). In order to guarantee that exposure to electromagnetic fields from wireless devices does not cause health hazards, many nations have established legislation or guidelines addressing SAR limits. These restrictions differ by area and are usually based on guidelines from global health organizations like the US Federal Communications Commission (FCC) and the International Commission on Non-Ionizing Radiation Protection (ICNIRP) [20]. A three-layer body tissue-equivalent phantom for skin, fat, and muscle layers of size 2,3, and 4 mm respectively is created as shown in Figure 7.5a and simulated in the Ansys HFSS tool. The overall size of the body phantom is $65 \times 65 \times 9\,mm^3$ and is considered to evaluate the SAR values over a tissue of 1 and 10 g. The comparison of $S_{11}$ characteristics and radiation pattern for free space and over the phantom is presented in Figure 7.5b–c.

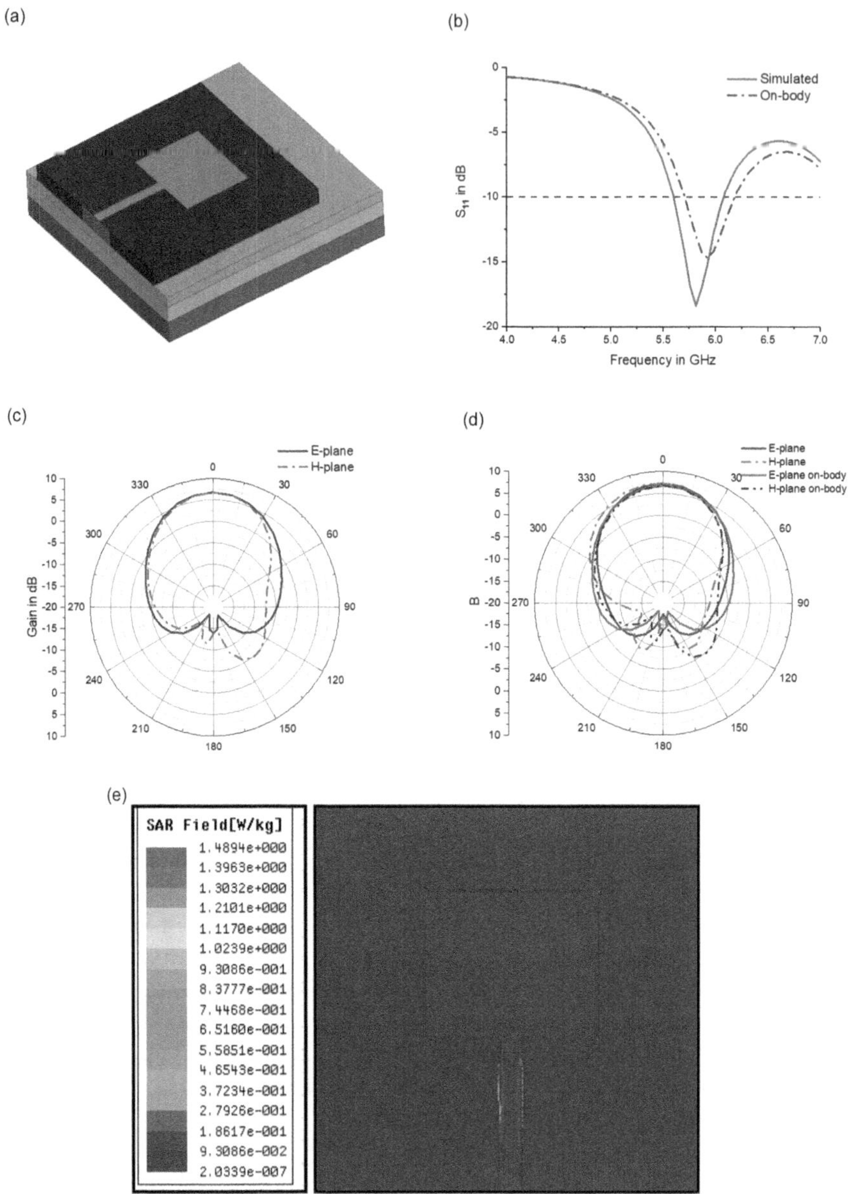


*Figure 7.5* Simulated results for SAR over a three-layer tissue equivalent phantom.

Table 7.3 shows the electrical parameters considered for the simulation of a three-layer tissue body phantom, and Table 7.4 depicts the value of $S_{11}$, bandwidth and gain for the comparison between on-body and off-body

*Table 7.3* Electrical characteristics of human tissue [21]

| *Parameter* | *Conductivity (S/m)* | *Relative permittivity* | *Loss tangent* | *Mass density* |
|---|---|---|---|---|
| Skin | 3.717 | 35.114 | 0.32807 | 1,109 |
| Fat | 0.29313 | 4.9549 | 0.18335 | 911 |
| Muscle | 4.9615 | 48.485 | 0.31715 | 1,090 |

*Table 7.4* Comparison of simulated results for on-body and off-body

| *Parameter* | *Resonating frequency (GHz)* | *$S_{11}$ (dB)* | *Bandwidth (MHz)* | *Gain (dB)* |
|---|---|---|---|---|
| Off-body | 5.8 | −18.75 | 470 | 7.2 |
| On-body | 5.9 | −14.91 | 468 | 6.77 |

scenarios. Figure 7.5e represents the SAR estimation for 1 and 10-g tissue. The maximum SAR obtained is 1.48 W/kg which is less than the safety limit prescribed by IEEE.

## 7.5 OPTIMIZATION IN WEARABLE ANTENNA

Optimizing wearable antennas involves enhancing their performance, efficiency, and integration into wearable devices while addressing constraints such as size, weight, and flexibility [22]. Here are some key steps and considerations for optimizing wearable antennas:

1. **Frequency Selection:** Choose the appropriate frequency band based on the application requirements. Higher frequencies offer wider bandwidth but may have shorter range and poorer penetration through materials, while lower frequencies offer better penetration but may require larger antennas.
2. **Antenna Design:** Design the antenna to match the specific requirements of the wearable device and application. Consider factors such as radiation pattern, impedance matching, polarization, and bandwidth. Compact and low-profile antenna designs are often preferred for wearable applications.
3. **Material Selection:** Choose materials with suitable electrical and mechanical properties for the antenna substrate and conductive elements. Flexible and lightweight materials are desirable for integration into wearable fabrics or structures.
4. **Integration with Wearable Devices:** Integrate the antenna seamlessly into the wearable device while minimizing its impact on form factor, comfort, and aesthetics. Position the antenna to optimize radiation

efficiency and minimize interference from the user's body or nearby objects.

5. **Matching Network Design**: Use impedance matching techniques such as matching networks or tuning components to ensure optimal power transfer between the antenna and the transmitter/receiver circuitry. This maximizes efficiency and improves overall performance.
6. **Performance Testing and Optimization**: Utilize simulation tools, prototyping, and testing to evaluate the antenna's performance in real-world conditions. Fine-tune the design parameters based on measured performance metrics such as return loss, radiation efficiency, and gain.
7. **Environmental Considerations**: Assess the antenna's performance under various environmental conditions encountered during typical use, such as humidity, temperature, and movement. Optimize the antenna design to maintain consistent performance across different scenarios.
8. **Multi-band and Wideband Operation**: Consider designing antennas capable of operating across multiple frequency bands or with wideband characteristics to support diverse wireless communication standards and future-proof the wearable device.
9. **User Interaction and Comfort**: Take into account user comfort and interaction when designing wearable antennas. Avoid sharp edges or protrusions that could cause discomfort or irritation, and ensure that the antenna does not restrict movement or hinder the user's activities.
10. **Regulatory Compliance**: Ensure that the wearable antenna complies with regulatory standards and SAR limits for human exposure to electromagnetic fields. Conduct SAR testing and certification as required by applicable regulations.

Through meticulous consideration of these optimisation stages and the utilisation of cutting-edge design methodologies and materials, engineers may create wearable antennas that satisfy the demands of contemporary wearable devices concerning performance, integration, and user experience.

## 7.6 CONCLUSION

A foam-based flexible wearable MSA operation in the ISM band at 5.8 GHz for healthcare applications is presented in this chapter. The designed geometry measures $50 \times 54 \times 3\,\text{mm}^3$ in size and resonates very well at 5.8 GHz, offering a maximum bandwidth of 470 MHz. The total gain obtained with the proposed antenna is 7.2 dBi. Unidirectional radiation is observed in both the *E* and *H* planes. SAR analysis is also carried out by structuring a three-layer body tissue equivalent phantom model of size $65 \times 65 \times 9\,\text{mm}^3$ with skin, fat, and muscle layers. The average SAR value for both 1 and 10-g tissue is 1.489 W/kg.

## REFERENCES

1. C. A. Balanis, *Balanis: Antenna Theory: Analysis and Design*. New York: John Wiley Sons Inc, 2001.
2. F. Meng and S. Sharma, A Single Feed Dual-Band (2.4 GHz/5 GHz) Miniaturized Patch Antenna for Wireless Local Area Network (WLAN) Communications, *Journal of Electromagnetic Waves and Applications*, vol. 30, no. 18, pp. 2390–2401, 2016.
3. P. Xu, M. Li, S. Wang, Y. Zhou, C. Shen, and X. Li, A Compact Multiband Antenna based on Metamaterial for WLAN/WiMAX/WAVE Applications, *APCAP*, 2017, Xi'an, China.
4. K. V. Karad and V. S. Hendre, Review of Antenna Array for 5G Technology Using mmWave Massive MIMO, *Lecture Notes in Electrical Engineering*, Singapore: Springer, 2022, pp. 775–787.
5. M. Kalyan and P. P. Sarkar, Dual Band Compact Monopole Antenna for ISM 2.4/5.8 Frequency Bands with Bluetooth, Wi-Fi and Mobile Applications, *Microwave and Optical Technology Letters*, vol. 59, pp. 1061–1065, 2017.
6. M. A. R. Osman, M. K. Abd Rahim, N. A. Samsuri, H. A. M. Salim, and M. F. Ali, "Embroidered Fully Textile Wearable Antenna for Medical Monitoring Applications, *Electromagnetic Waves*, vol. 117, pp. 321–337, 2011.
7. J. Xinghui, M. Zhu, W. Wei, G. Yu, and H. Zho, A Small-Size and Multi-Band Wearable Antenna with Coplanar Structure, *Journal of Physics: Conference Series*, vol. 1325, no. 1, pp. 278–285, 2019.
8. S. N. Mahmood et al., Recent Advances in Wearable Antenna Technologies: A Review, *Progress in Electromagnetics Research B (PIER B)*, vol. 89, pp. 1–27, 2020.
9. D. Mamta, A. Sharma, S. Yadav, and R. Singhal, Performance Investigation of Flexible UWB Antenna Near Human Body for Wearable Appliances, *Progress in Electromagnetics Research B*, vol. 97, pp. 131–147, 2022.
10. U. Ali, S. Ullah, B. Kamal, L. Matekovits, and A. Altaf, Design, Analysis and Applications of Wearable Antennas: A Review, *IEEE Access*, vol. 11, pp. 14458–14486, 2023.
11. A. Gupta, A. Kansal, and P. Chawla, A Survey and Classification on Applications of Antenna in Health Care Domain: Data Transmission, Diagnosis and Treatment, *Sadhana*, vol. 46, no. 2, pp. 158–165, 2021.
12. K. V. Karad and V. S. Hendre, A Foam-Based Compact Flexible Wideband Antenna for Healthcare Applications, *Progress in Electromagnetics Research C (PIER C)*, vol. 123, pp. 197–212, 2022.
13. L. Yang et al., Review on Wearable Antenna Design, *Lecture Notes in Electrical Engineering*, Singapore: Springer, 2021, pp. 731–742.
14. G. J. Hayes, J.-H. So, A. Qusba, M. D. Dickey, and G. Lazzi, Flexible Liquid Metal Alloy (EGaIn) Microstrip Patch Antenna, *IEEE Transactions on Antennas and Propagation*, vol. 60, no. 5, pp. 2151–2156, 2012.
15. H. R. Raad, A. I. Abbosh, H. M. Al-Rizzo, and D. G. Rucker, Flexible and Compact AMC Based Antenna for Telemedicine Applications, *IEEE Transactions on Antennas and Propagation*, vol. 61, no. 2, pp. 524–531, 2013.

16. R. B. V. B. Simorangkir, Y. Yang, R. M. Hashmi, T. Bjorninen, K. P. Esselle, and L. Ukkonen, Polydimethylsiloxane-Embedded Conductive Fabric: Characterization and Application for Realization of Robust Passive and Active Flexible Wearable Antennas, *IEEE Access*, vol. 6, pp. 48102–48112, 2018.
17. B. V. B. Roy, S. Simorangkir, Y. Ieee, L. Ieee, S. Matekovits, and K. P. Ieee, Dual-Band Dual-Mode Textile Antenna on PDMS Substrate for Body-Centric Communications, *IEEE Antennas and Wireless Propagation Letters*, vol. 2872, no. 12, pp. 220–228, 2016.
18. K. V. Karad and V. S. Hendre, Design and Simulation of 28 GHz Antenna for 5G mmWave Communication, *Lecture Notes in Electrical Engineering*, Singapore: Springer Nature, 2022, pp. 885–890.
19. P. M. Rayner and W. G. Whittow, Specific Absorption Rate and Efficiency of a Wideband Wearable Monopole Antenna Near the Human Body, *Loughborough Antennas & Propagation Conference (LAPC 2017)*, Loughborough, UK, 2017.
20. S. Doddipalli, A. Kothari, and P. Peshwe, A Low Profile Ultrawide Band Monopole Antenna for Wearable Applications, *International Journal of Antennas and Propagation*, vol. 2017, p. 2249, 2017.
21. Dielectric Properties of Body Tissues: HTML Clients, Cnr.it. [Online]. Available: https://niremf.ifac.cnr.it/tissprop/htmlclie/htmlclie.php. [Accessed: 25-Feb-2024].
22. Ł. Januszkiewicz, P. D. Barba, and J. Kawecki, Design Optimization of Wearable Multiband Antenna Using Evolutionary Algorithm Tuned with Dipole Benchmark Problem, *Electronics (Basel)*, vol. 10, no. 18, p. 2249, 2021.

Chapter 8

# Bandwidth and gain enhancement of a slotted S-shaped microstrip patch antenna for 5G application

*Dipankar Saha, Sudip Mandal, and Kabita Purkait*

## 8.1 INTRODUCTION

The rapid advancement of wireless communication systems from 1G to 5G has led to higher data rates and greater connectivity. Due to the overcrowded spectrum of 4G and the demand for higher data speed, it is replaced with 5G technology. 5G promises to provide wider bandwidth, greater connectivity, low latency and good data rates. IMT2020 is the authority term for 5G in ITU-R particulars. IMT2020 introduced 5G recurrence groups somewhere in the range of 24 and 71 GHz with a complete range transmission capacity designation of 17.25 GHz [1]. Rappaport et al. [2] concluded that 28 and 38 GHz frequencies for 5G are ideal for enforcing future mobile communication networks. With the progressive approach of wireless communication, a simple and compact antenna is needed to able with 5G technology. The microstrip patch antenna has become very popular over the years due to its compactness, ease of fabrication, low cost, and light weight. However, it also has some drawbacks, including limited bandwidth, lower efficiency and reduced gain.

A conventional microstrip patch antenna consists of three components: the substrate, the patch and the ground. The metallic patch is placed on the dielectric material and upheld by the ground plane. The relative permittivity of the substrate ($\varepsilon_r$) varies from $2.2 \leq \varepsilon_r \leq 12$ to design a microstrip antenna. The patch can be made up of copper or gold and may have many shapes, including rectangle, square, triangle, circular, elliptical etc. [3]. The substrate's relative permittivity regulates the antenna's gain and bandwidth. The size of the antenna is also controlled by the substrate. If antenna size decreases, bandwidth also decreases. The suggested antenna is configured using an FR-4 substrate whose relative permittivity is 4.4 and loss tangent is 0.025. The suggested microstrip patch antenna's substrate materials have a top height of 4 mm ($h$).

The bandwidth of the antenna can be enhanced by using various techniques like loading slots, slits, notches in the patch or ground. Marotkar and Zade [4] exhibited a rectangular microstrip patch antenna with defected ground

DOI: 10.1201/9781003560487-8

configuration having multiple slots which provides bandwidth enhancement from 67 to 302 MHz. Zakir Ali et al. [5] proposed bandwidth enhancement of an E-shaped microstrip patch antenna to 14.90%. Diwakar Singh et al. [6] proposed an S-shaped antenna for Wi-max application. Enhancement of bandwidth achieved is 28.0056% at 2.2382 GHz. Menaka et al. [7] presented an S-shaped microstrip patch antenna with a bandwidth enhancement of 70 MHz at 4.58 GHz. Ali Abdulateef Abdulbari et al. [8] designed a T-shaped antenna made up of Rogers RT 588 lz substrate with bandwidth enhancement of 42.81% at 3.6 GHz. Dian Widi Astuti et al. [9] presented a bow-tie microstrip antenna using FR-4 with a bandwidth increase from 2.6% to 10.45% at 3.67 GHz. A circular patch was included as a parasitic element in Tang et al.'s proposed wideband strip-helical antenna [10], which increased the impedance bandwidth to 56%. A T-shape patch antenna with a bandwidth increase from 40.05% to 81.34% in the frequency range of 1.682–3.988 GHz was proposed by Ramesh Kumar Verma and Srivastava [11] utilizing FR-4 substrate. WeiXing Liu et al. [12] presented a rectangular microstrip patch antenna with a U-shaped slot that provides bandwidth enhanced from 28% to 122%. Wang and Lancaster [13] proposed a different approach to bandwidth enhancement by using a meander patch antenna coupled by an H-shaped aperture by changing the antenna's electrical route length. An impedance bandwidth of 1.72 GHz was achieved by Mohammed et al. [14] through mathematical analysis and optimization of the impact of increasing conductor thickness on microstrip patch antenna performance. Kumar and Modani [15] have presented a rectangular patch antenna with a 1.1 GHz bandwidth, a return loss of –18.25 dB, and a gain of 6.83 dBi. Planar inverted-F antenna at 28 GHz with 1.55 GHz of bandwidth and 4.47 dB gain is shown by Waleed Ahmad et al. [16].

In this chapter, a rectangular microstrip patch antenna is initially designed with a notch at the lower end of the left-side patch antenna. By creating a vertical slit beneath the first arm of the S-shaped patch antenna and cutting a notch in the right end of the patch antenna to produce an S, the bandwidth of the antenna is increased. Finally, another slot is cut out horizontally to enhance the overall bandwidth of the S-shape microstrip patch antenna. The presented antenna is feeding with a 50 Ω microstrip line. An S-shape patch antenna is proposed using remodelling the rectangular patch at the desired frequency of 28 GHz and all simulation work is done in the CST studio environment [17].

## 8.2 METHODOLOGY

By using the following formulas, one may determine the antenna's length and width [3]:

$$W = \frac{C}{2f_r}\sqrt{\frac{2}{\varepsilon_r + 1}} \tag{8.1}$$

where $C$ is speed of light, $f_r$ is the resonant frequency, $\varepsilon_r$ is the relative permittivity of the substrate

$$L = L_{eff} - 2\Delta L \tag{8.2}$$

where $L_{eff}$ is the effective length of the patch can be expressed as

$$L_{eff} = \frac{c}{2f_r\sqrt{\epsilon_{eff}}} \tag{8.3}$$

where $\epsilon_{eff}$ is the effective relative permittivity given as follows:

$$\epsilon_{eff} = \frac{\epsilon_r + 1}{2} + \frac{\epsilon_r - 1}{2}\left(1 + \frac{12h}{W}\right)^{-\frac{1}{2}} \tag{8.4}$$

and $\Delta L$ is the length extension given as follows:

$$L = 0.412h \frac{(\epsilon_{eff} + 0.3)\left(\frac{W}{h} + 0.264\right)}{(\epsilon_{eff} - 0.258)\left(\frac{W}{h} + 0.8\right)} \tag{8.5}$$

The dimensions (length×width×height) of the patch antenna are approximately indicated by the corresponding length $L_g$, width $W_g$ and height of the substrate $h$. The provided formulae can be used to determine these values.

$$L_g = L + 6h \tag{8.6}$$

$$W_g = W + 6h \tag{8.7}$$

## 8.3 ANTENNA DESIGN SPECIFICATION

FR-4 substrate with a relative permittivity of 4.4 is used in the construction of the suggested S-shaped microstrip patch antenna at the intended frequency of 28 GHz. The metal patch's thickness is assumed to be 0.035 mm. Table 8.1 lists the various specifications for the suggested microstrip patch antenna. The microstrip antenna is fed via a 50 Ω Microstrip line.

## 8.4 ANTENNA DESIGN PROCEDURE

The design of a microstrip patch antenna relies on the selection of frequency, the choice of proper substrate material and the dimensions of the patch, all of which play a very vital role in structuring the antenna.

*Table 8.1* The parameters for the proposed microstrip patch antenna

| *Parameters* | *Values (mm)* |
|---|---|
| Width (*W*) | 4.85 |
| Length (*L*) | 3.6 |
| Width of ground ($W_g$) | 7.2 |
| Length of ground ($L_g$) | 7 |
| Width of feed ($W_f$) | 1 |
| Length of feed ($L_f$) | 1.8 |
| Height of Substrate (*h*) | .40 |

A rectangular microstrip patch antenna with size of 7.2 mm×7 mm×40 mm using FR-4 substrate is designed initially and implemented using CST studio suite. The desired frequency for the proposed modelling is 28 GHz. The proposed antenna's frequency range, as determined through simulation of the design, is between 26.676 and 28.397 GHz. The 50 Ω microstrip feedline is carried out in the middle of the patch. Figure 8.1 depicts the antenna geometry design process step by step, and Table 8.2 lists the slot and notch parameters.

As shown in Figure 8.1a, antenna I is first created in the first step by creating a notch at the patch's left lower boundary that measures 2.43×0.70 mm. The bandwidth of the first phase antenna is 3.011%, with a frequency range between 27.507 and 28.348 GHz and a return loss of –20.39 dB. In the subsequent phase, the patch is cut with a further notch measuring 2.22 mm×0.90 mm and a slot measuring 1.80 mm× 0.60 mm. The bandwidth is reached at 4.59% between the frequency band of 26.627 and 27.878 GHz with a reflection coefficient of –31.05 dB. Lastly, the left centre of the patch is sliced with a notch measuring 2.20 mm×0.50 mm. The antenna's bandwidth is acquired 6.485% between the frequency band of 26.614 and 28.398 GHz, resonating at 27.524 GHz with a significant return loss of –40.40 dB. In Table 8.3, a comparative result of different antennas is displayed.

## 8.5 RESULT AND DISCUSSION

A slotted S-shaped microstrip patch is designed to enhance the bandwidth to 6.485%. Since the original design of the microstrip patch antenna had poor bandwidth and gain, the aim of this work is to increase these parameters. All simulations and analyses for designing this microstrip patch antenna were executed using CST studio suite software. From 25 to 31 GHz, the suggested antenna is experimented. The last slotted S-shaped microstrip patch antenna is demonstrated at a resonant frequency of 27.524 GHz with

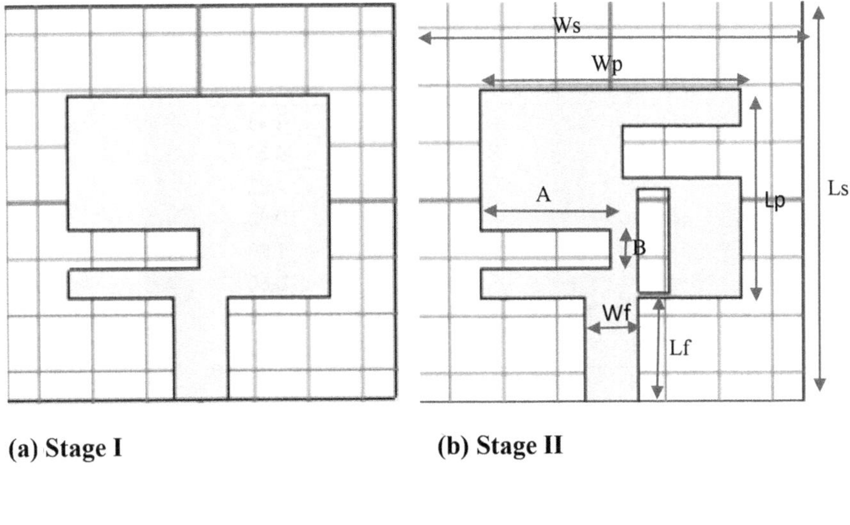


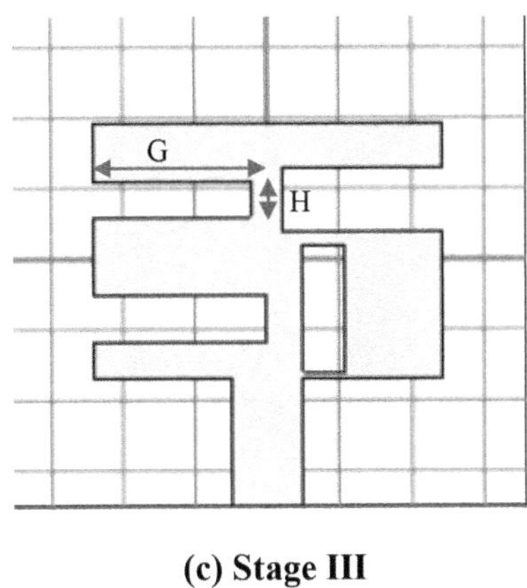


*Figure 8.1* The geometry of three different stages of the proposed slotted antenna. (a) Stage I, (b) Stage II, and (c) Stage III.

achieved bandwidth enhancement of 6.485% and return loss of –40.40 dB. The return loss of three different antennas is compared, as shown in Figure 8.2. Return loss ($S_{11}$) indicates how much power is reflected back from load to source. Antenna stage I provide –20.39 dB of return loss while antenna stage II exhibits –31.05 dB of return loss and antenna stage III yields excellent return loss of –40.40 dB. Three antennas' VSWRs are also contrasted, as seen in Figure 8.3. VSWR suggests how much the antenna matches with the transmission line. Antenna stage I have a VSWR of 1.21, while antenna stage II produces a VSWR of 1.05 and antenna stage III registers

*Table 8.2* The parameters for slotted S-shaped microstrip patch antenna

| *Parameters* | *Values (mm)* |
|---|---|
| *A* | 2.43 |
| *B* | 0.70 |
| *C* | 2.22 |
| *D* | 0.90 |
| *E* | 1.80 |
| *F* | 0.60 |
| *G* | 2.20 |
| *H* | 0.50 |

*Table 8.3* Contrastive outcomes of different antenna modelling

| *Antenna stages* | *Resonant frequency (GHz)* | *Return loss(dB)* | *Bandwidth (%)* |
|---|---|---|---|
| Stage I | 27.926 | −20.39 | 3.011 |
| Stage II | 27.159 | −31.05 | 4.59 |
| Stage III | 27.524 | −40.4 | 6.485 |

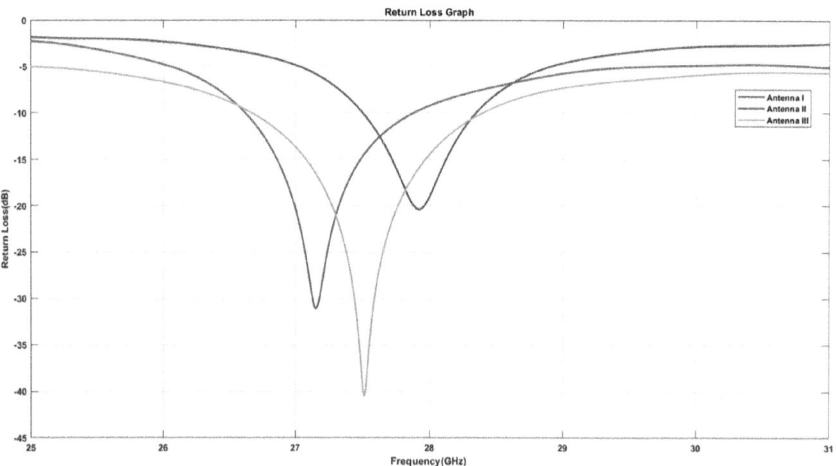


*Figure 8.2* Graph of return loss against frequency.

itself good matching device with a VSWR of 1.01. Figure 8.4 represents the comparative gain graph among three different antennas. Antenna stage III outperforms the other antennas and provides a good gain of 7.602 dB. The directivity of the slotted S-shaped microstrip patch antenna (antenna stage III) is 8.078, as shown in Figure 8.5. Figure 8.6 shows that at the

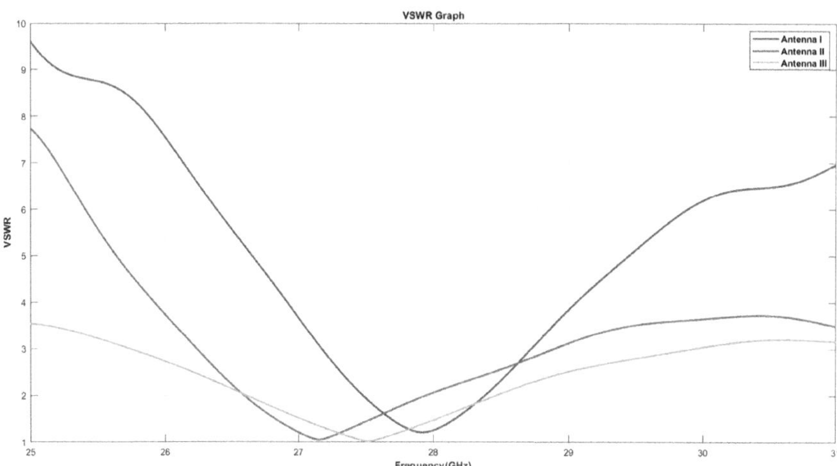


*Figure 8.3* VSWR versus frequency graph.

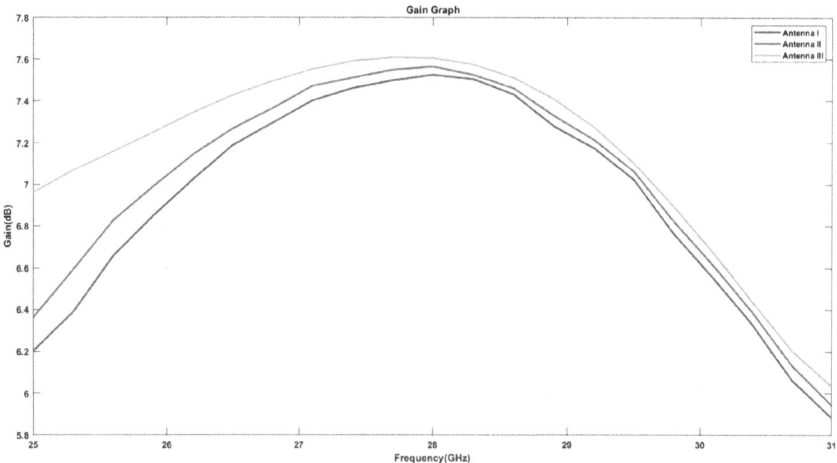


*Figure 8.4* Gain versus frequency graph.

resonance frequency of 27.524 GHz, antenna stage III achieves an efficiency of 89.60%. Figures 8.7–8.9 depict 2D and 3D radiation characteristics of slotted S-shaped microstrip patch antenna at $\varphi=0°$ and $\varphi=90°$ respectively and 3 dB angular width at the resonant frequency of 27.524 is 55.2° and 145°. Surface Current circulation of the slotted antenna stage III is shown in Figure 8.10.

The outcomes of the slotted S-shaped antenna stage III are compared with previous findings. Table 8.4 compares the suggested antenna with the

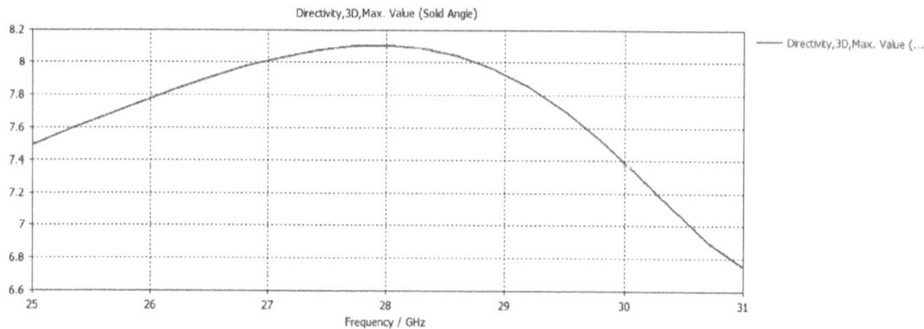


*Figure 8.5* Directivity of the slotted antenna stage III against frequency.

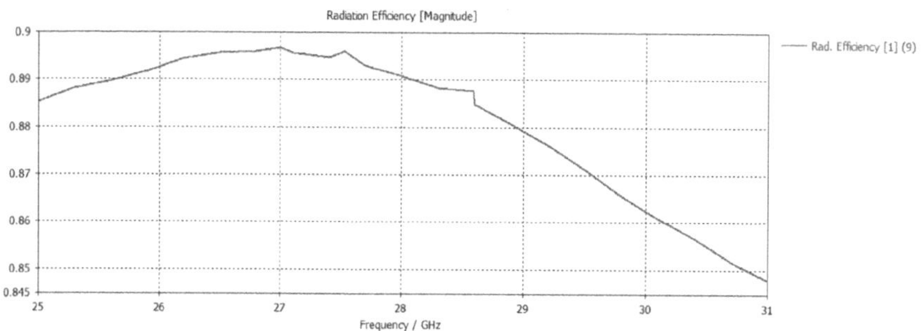


*Figure 8.6* Efficiency of the slotted antenna stage III against frequency.

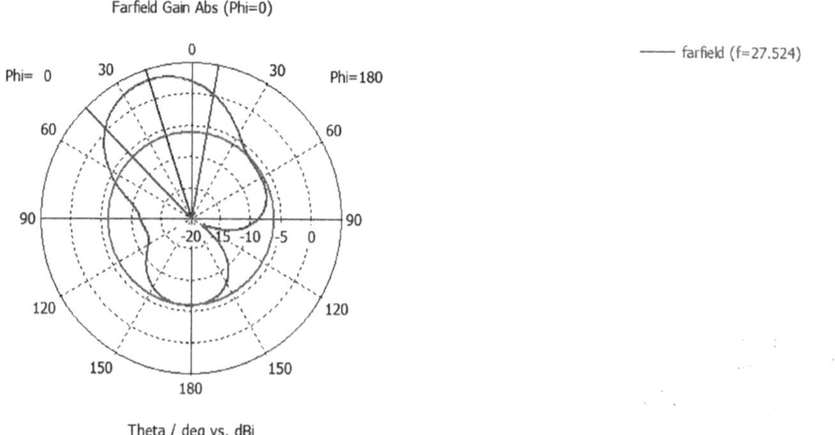


*Figure 8.7* The radiation pattern of the slotted antenna stage III when $\varphi = 0°$.

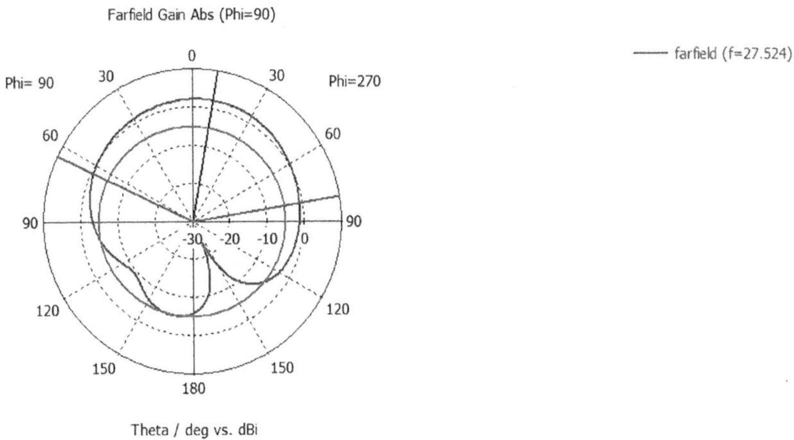


*Figure 8.8* The radiation pattern of the slotted antenna III when $\varphi=90°$.

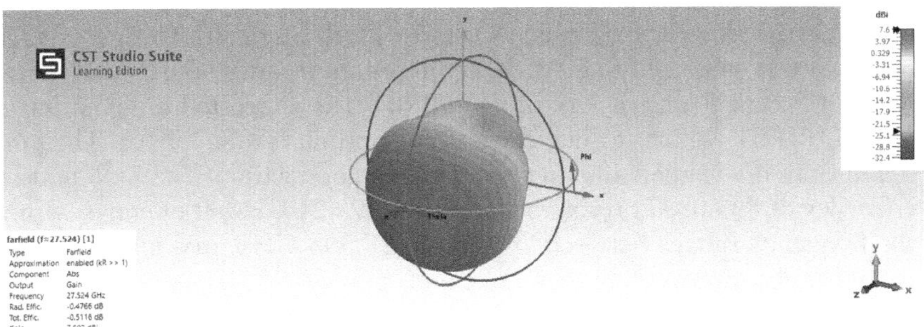


*Figure 8.9* 3D slotted antenna stage III radiation pattern.

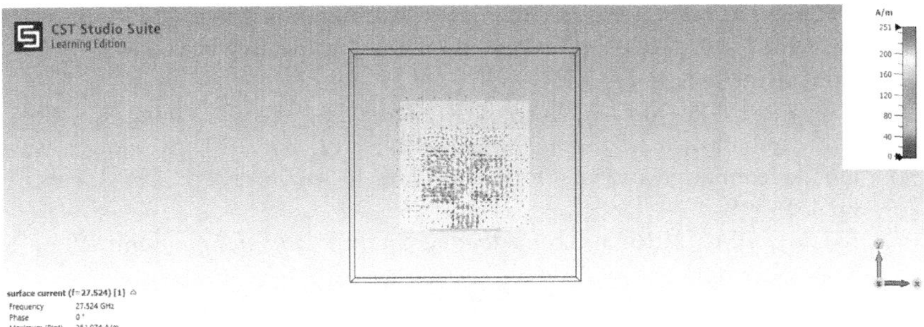


*Figure 8.10* The slotted antenna stage III's surface current.

*Table 8.4* Contrasts previous patch antenna designs with the recommended patch antenna design

| *References* | *Resonant frequency (GHz)* | *Size (mm²)* | *Return loss (dB)* | *Bandwidth (GHz)* | *Gain (dBi)* |
|---|---|---|---|---|---|
| [14] | 28 | 5.34×4.54 | 39.49 | 1.72 | 9.55 |
| [15] | 28 | 4.74×3.45 | −18.25 | 1.1 | 6.83 |
| [16] | 28 | 4.40×4.20 | Not Specified | 1.55 | 4.47 |
| Proposed Work | 27.524 | 4.85×3.6 | −40.40 | 1.78 | 7.6 |

alternative with respect to bandwidth, resonant frequency, gain, and return loss. From comparative results, the suggested patch antenna provides good bandwidth, high gain and excellent return loss.

## 8.6 CONCLUSION

Using various notches and slots, a rectangular patch antenna is redesigned to create the slotted S-shaped microstrip patch antenna. The percentage bandwidth is enhanced to 6.485% at a resonant frequency of 27.524 GHz, with an excellent return loss of –40.40 dB. The suggested antenna has a VSWR of 1.01, which is quite close to the optimum value of 1.0. The proposed slotted S-shaped antenna also yields a good gain of 7.602 dB and an efficiency of 89.60% at resonant frequency 27.524 GHz. It operates within the frequency ranges between 26.614and 28.398 GHz, making it suitable for 5G applications.

## REFERENCES


1. International Telecommunication Union– Radiocommunication Sector (ITU-R) FAQ on International Mobile Telecommunications (IMT). Accessed on: April 30, 2020. [Online]. Available: https://www.itu.int/en/ITU-R/Documents/ITU-R-FAQ-IMT.pdf.
2. Rappaport, T.S., Sun, S., Mayzus, R., Zhao, H., Azar, Y., Wang, K., Wong, G.N., Schulz, J.K., Samimi, M. and Gutierrez, F., 2013. Millimeter wave mobile communications for 5G cellular: It will work! *IEEE Access*, 1, pp.335–349.
3. Balanis, C.A., 2016. *Antenna theory: analysis and design*. John Wiley & Sons, New York.
4. Marotkar, D.S. and Zade, P., 2016, March. Bandwidth enhancement of microstrip patch antenna using defected ground structure. *2016 International conference on electrical, electronics, and optimization techniques (ICEEOT)* (pp. 1712–1716). IEEE, Chennai.

5. Ali, Z., Singh, V.K., Singh, A.K. and Ayub, S., 2011, October. E-Shaped microstrip antenna on Rogers substrate for WLAN applications. *2011 International conference on computational intelligence and communication networks* (pp. 342–345). IEEE, Gwalior.
6. Singh, D., Gupta, A.K. and Prasad, R.K., 2013. Analysis and design of S-shape microstrip patch antenna. *Journal of Electronics and Communication Engineering*, 7(4), pp.18–22.
7. Menaka, R., Nishandhi, S. and Sivaranjani, S., 2017. Designing of S shaped microstrip patch antenna for broadband application using slotting technique. *IJSR*, 6(3), pp.2284–2287.
8. Abdulbari, A.A., Jawad, M.M., Hanoosh, H.O., Saare, M.A., Lashari, S.A., Sari, S.A., Ahmad, S., Khalill, Y. and Hussain, Y.M., 2021. Design compact microstrap patch antenna with T-shaped 5G application. *Bulletin of Electrical Engineering and Informatics*, 10(4), pp.2072–2078.
9. Astuti, D.W., Fadilah, R., Muslim, D.R., Rusdiyanto, D., Alam, S. and Wahyu, Y., 2022. Bandwidth enhancement of bow-tie microstrip patch antenna using defected ground structure for 5G. *Journal of Communication*, 17(12), pp.995–1002.
10. Tang, X., He, Y. and Feng, B., 2016. Design of a wideband circularly polarized strip-helical antenna with a parasitic patch. *IEEE Access*, 4, pp.7728–7735.
11. Verma, R.K. and Srivastava, D.K., 2019. Bandwidth enhancement of a slot loaded T-shape patch antenna. *Journal of Computational Electronics*, 18, pp.205–210.
12. Liu, W.X., Yin, Y.Z., Xu, W.L. and Zuo, S.L., 2011. Compact open-slot antenna with bandwidth enhancement. *IEEE Antennas and Wireless Propagation Letters* 10, pp.850–853.
13. Wang, H.Y. and Lancaster, M.J., 1999. Aperture-coupled thin-film superconducting meander antennas. *IEEE Transactions on Antennas and Propagation*, 47(5), pp.829–836.
14. A.S. Mohammed, S. Kamal, M.F.B. Ain, R. Hussin, F. Najmi, S.A. Sundi, Z.A. Ahmad, U. Ullah, M.F.B.M. Omar, M. Othman, et al., 2021. Mathematical model on the effects of conductor thickness on the centre frequency at 28 GHz for the performance of microstrip patch antenna using air substrate for 5G application, *Alexandria Engineering Journal*, 60(6), pp.5265–5273.
15. Kumar, G.R. and Modani, U. S., 2018. A compact microstrip patch antenna at 28 GHz for 5G wireless applications. *2018 3rd International conference and workshops on recent advances and innovations in engineering (ICRAIE)* (pp. 1–2). IEEE, Rajasthan, India.
16. Ahmad, W., Ali, A., and Khan, W. T., 2016, June. Small form factor PIFA antenna design at 28 GHz for 5G applications. *2016 IEEE international symposium on antennas and propagation (APSURSI)* (pp. 1747–1748). IEEE, Fajardo, Puerto Rico, USA.
17. CST Studio Suite, Available: https://edu.3ds.com/en/software/cst-studio-suite-students-edition. Accessed on: July 05, 2022.

Chapter 9

# Advancements in biomedical antennas

## Exploring diverse designs for wireless connectivity in medical applications

*N. Kavitha and P. Palaniyammal*

## 9.1 INTRODUCTION

A biomedical antenna with optimization techniques for wireless communication is a specialized component meticulously designed for applications within the healthcare and medical technology domains. Unlike conventional antennas, biomedical antennas are tailored to operate in or around the human body, addressing the intricate challenges associated with wireless communication in these environments. Optimization techniques are applied to augment the antenna's performance, efficiency, and reliability, ensuring seamless communication in healthcare applications [1–5].

One key consideration in the optimization process is biocompatibility. Biomedical antennas must interact with the human body or medical implants without causing harm, and optimization strategies focus on selecting materials and designs that are safe and non-intrusive. Additionally, miniaturization is crucial, especially for integration into small medical devices or wearables. Optimization techniques aim to reduce the size of the antenna while preserving or enhancing its functionality.

Efficient power consumption is another critical aspect of optimization, particularly for devices powered by batteries or energy-harvesting systems. Optimization techniques work toward minimizing power requirements, ultimately extending the battery life of implantable medical devices. The frequency and bandwidth of the antenna are also optimized to align with the specific requirements of biomedical applications, ensuring effective data transmission within the desired frequency bands.

Optimization encompasses the strategic placement of the antenna, considering factors such as the human body's composition and movements. The goal is to position the antenna optimally to maximize signal strength, minimize interference, and adapt to changes in the body's conditions or movements. Signal integrity is prioritized through optimization, aiming to maintain reliable communication for accurate data transmission.

In some instances, optimization involves the use of multiple antennas in a system (Multiple Input Multiple Output systems). This approach enhances

 DOI: 10.1201/9781003560487-9

signal diversity, reliability, and overall communication performance. Frequency reconfigurability may also be integrated, allowing the antenna to adapt to different communication requirements or environmental conditions.

Material selection is a crucial component of optimization. The chosen materials should not only facilitate efficient signal transmission but also comply with safety standards for medical use. By carefully considering these factors, biomedical antennas with optimization techniques play a pivotal role in advancing wireless communication within healthcare applications. They enable vital functions such as remote patient monitoring, implantable medical device communication, telemedicine, and the seamless integration of wearable health technologies, ultimately contributing to improved patient care and healthcare outcomes [6–10].

## 9.2 TYPES OF BIOMEDICAL ANTENNAS

Biomedical antennas are specialized components designed for wireless communication within or around the human body, as well as for various medical applications. These antennas play a crucial role in facilitating communication in healthcare devices, including implantable medical devices, wearables, and body area networks (BANs). Below are detailed explanations of some types of biomedical antennas:

### 9.2.1 Microstrip antennas

Microstrip antennas are characterized by their planar configuration, featuring a metal strip on a dielectric substrate. Notable for their compact size and low profile, they are well-suited for integration into small medical devices or wearables, offering an efficient solution for wireless communication.

### 9.2.2 Dipole antennas

Dipole antennas consist of a straight conductor divided into two halves, forming poles. Simple yet effective, they find application in short-range medical telemetry and BANs, serving as reliable components for communication within the human body.

### 9.2.3 Patch antennas

Patch antennas, with a flat and often rectangular design, are printed or etched on a substrate. Their compact structure and ease of integration make them suitable for implantable devices and wearable health monitors, contributing to the miniaturization of medical devices.

### 9.2.4 Helical antennas

Helical antennas feature a coil shape and provide circular polarization. They are particularly useful in applications demanding robust communication with changing orientations, such as wireless capsule endoscopy (WCE), demonstrating efficacy in medical scenarios where polarization consistency is crucial.

### 9.2.5 Inverted-F antennas

Inverted-F antennas are planar and compact, combining features of monopole and loop antennas. Their versatility and ease of integration make them suitable for wireless healthcare monitoring systems, contributing to reliable communication in various medical contexts.

### 9.2.6 Slot antennas

Slot antennas have openings in the conducting surface, offering a solution for integration into medical implants and devices with limited space. This design is tailored to meet the constraints of compact medical applications while maintaining efficient wireless communication.

### 9.2.7 Planar inverted-F antennas (PIFA)

PIFA antennas integrate advantages from both microstrip and inverted-F antennas. Their planar design and versatility make them suitable for implantable devices and portable medical equipment, demonstrating adaptability in various medical settings.

### 9.2.8 Wireless sensor antennas

Designed specifically for wireless sensor networks in healthcare, these antennas contribute to remote patient monitoring and data transmission in medical applications, supporting the advancement of sensor technologies in healthcare systems.

### 9.2.9 Ultra-wideband (UWB) antennas

UWB antennas transmit over a wide frequency range, facilitating high-data-rate communication. Their applications in medical imaging and diagnostics enable precise localization and imaging, crucial for advanced healthcare technologies.

### 9.2.10 Circularly polarized antennas

Circularly polarized antennas, known for producing circularly polarized waves, are employed in medical imaging for improved signal reception. Their application in healthcare scenarios contributes to enhanced signal integrity and reliability.

### 9.2.11 Conformal antennas

Conformal antennas are designed to conform to irregular surfaces, making them suitable for integration into medical devices with non-standard shapes. This adaptability ensures efficient wireless communication without compromising device form.

### 9.2.12 Dielectric resonator antennas (DRA)

DRA antennas use dielectric materials to create resonant structures, serving applications in medical imaging where high-frequency performance is essential. Their design aligns with the stringent requirements of medical diagnostic equipment.

Incorporating these diverse types of biomedical antennas with specific designs and functionalities ensures a comprehensive approach to wireless communication in healthcare, addressing the unique challenges and requirements of medical applications.

## 9.3 MEANDERED ANTENNAS IN BIOMEDICAL APPLICATIONS

A meandered antenna is a specific type of antenna design that features a winding or serpentine path instead of a straight-line configuration. This meandering structure offers several advantages in terms of compactness, size reduction, and impedance matching, making it suitable for various applications, including biomedical applications. Meandered antennas, characterized by their serpentine or winding structure, have become integral components in various biomedical applications due to their unique design features. Their compact size, conformal integration capabilities, and suitability for multiband operation make them particularly well-suited for healthcare scenarios. One illustrative example of their application is in WCE, a non-invasive diagnostic technique used for imaging the gastrointestinal (GI) tract.

In WCE, patients ingest a small capsule equipped with a camera, light source, and wireless transmitter. As the capsule navigates the GI tract, it captures and wirelessly transmits images for diagnostic purposes.

Meandered antennas play a pivotal role in this context, offering advantages such as compact integration, conformal placement, multiband operation, biocompatibility, stability in dynamic environments, low power consumption, and potential for energy harvesting. These characteristics make meandered antennas crucial for real-time monitoring of the GI tract, showcasing their versatility and effectiveness in addressing the specific challenges posed by biomedical applications, particularly in the field of diagnostic imaging.

## 9.3.1 Wireless capsule endoscopy

WCE is a non-invasive diagnostic technique used for imaging the GI tract. In this procedure, patients swallow a small capsule containing a tiny camera, light source, and wireless transmitter. As the capsule traverses the digestive system, it captures images that are wirelessly transmitted to an external receiver worn by the patient.

### *9.3.1.1 Compact integration*

The limited space within the capsule necessitates a compact antenna design. Meandered antennas, with their serpentine structure, allow for efficient use of the available space, ensuring the antenna fits within the small confines of the capsule without compromising performance.

### *9.3.1.2 Conformal placement*

Meandered antennas can be conformally placed on the outer surface of the capsule. Their flexible design enables them to follow the contours of the capsule, optimizing the antenna's exposure to the external environment for reliable wireless communication.

### *9.3.1.3 Multiband operation*

Biomedical applications often involve the use of different wireless communication standards. Meandered antennas can be designed to operate across multiple frequency bands, ensuring compatibility with various communication protocols and standards used in WCE systems.

### *9.3.1.4 Biocompatibility*

Materials chosen for the meandered antenna can be selected for biocompatibility, ensuring that the antenna does not cause harm when in contact with the digestive tract. Biocompatible materials are crucial to prevent adverse reactions or irritation during the capsule's journey through the GI tract.

#### *9.3.1.5 Stability and resilience*

The meandered structure enhances the antenna's stability and resilience to detuning caused by changes in the surrounding biological environment. This is particularly important in the dynamic and variable conditions of the GI tract.

#### *9.3.1.6 Low power and energy harvesting*

Meandered antennas can be optimized for low-power communication, extending the operational life of the capsule's battery. Additionally, energy harvesting techniques, such as capturing ambient RF energy, can be integrated into the antenna design to further enhance the longevity of the capsule's power source.

#### *9.3.1.7 Real-time monitoring*

The wireless communication facilitated by meandered antennas allows real-time monitoring of the GI tract. Medical professionals can receive continuous updates on the condition of the digestive system, aiding in the diagnosis of GI disorders and diseases.

In summary, meandered antennas play a crucial role in WCE by providing a compact, conformal, and efficient means of wireless communication. Their design characteristics address the specific challenges presented by the confined and dynamic environment of the GI tract, showcasing the versatility and applicability of meandered antennas in biomedical applications.

## 9.4 PATCH ANTENNAS IN BIOMEDICAL APPLICATIONS

Patch antennas are planar antennas that consist of a conducting patch on a dielectric substrate, often backed by a ground plane. These antennas are widely used in various applications, including biomedical scenarios, due to their compact design, low profile, and ease of integration.

### 9.4.1 Wearable health monitoring devices

Wearable health monitoring devices have gained significant popularity for their ability to continuously track and monitor various physiological parameters, promoting proactive healthcare. Patch antennas, due to their compact size, low profile, and efficiency, are commonly integrated into these devices to enable wireless communication. Here's an overview of the use of patch antennas in wearable health monitoring devices:

#### 9.4.1.1 Application of patch antennas in wearable fitness trackers: real-time heart rate monitoring

Real-time heart rate monitoring stands as a pivotal application of patch antennas within wearable fitness trackers. This integration leverages wireless communication technology to offer users immediate insights into their cardiovascular activity, enhancing the overall health monitoring experience.

#### 9.4.1.2 Patch antenna integration

Wearable fitness trackers incorporate patch antennas within their internal structure. These antennas are strategically placed, often in proximity to heart rate sensors, to optimize wireless communication.

#### 9.4.1.3 Heart rate sensor operation

Heart rate sensors, utilizing optical or electrical technologies, measure physiological signals to capture real-time data indicative of the user's heart rate during physical activities.

#### 9.4.1.4 Wireless communication

Patch antennas play a crucial role in facilitating wireless communication within the fitness tracker. They transmit heart rate data wirelessly to an internal receiver or processing unit within the device.

#### 9.4.1.5 Real-time monitoring

The fitness tracker's user interface displays real-time heart rate information, providing users with immediate feedback on their cardiovascular activity. This enables on-the-fly adjustments to exercise intensity or pace.

#### 9.4.1.6 Wireless connectivity

Patch antennas enable seamless communication with external devices such as smartphones or smartwatches. Heart rate data is transmitted to dedicated fitness apps, offering users a comprehensive overview of their cardiovascular performance.

#### 9.4.1.7 Biocompatibility and comfort

Materials chosen for patch antennas prioritize biocompatibility to ensure safety and comfort during prolonged skin contact. Both the fitness tracker and the patch antenna are designed with user comfort in mind.

#### *9.4.1.8 Optimization for power efficiency*

Patch antenna designs are optimized for low power consumption. This ensures extended battery life, allowing users to utilize the fitness tracker for longer periods between charges.

#### *9.4.1.9 Data logging and analysis*

Fitness trackers log heart rate data for later analysis. Syncing with a user's smartphone app provides insights into heart rate trends, recovery patterns, and overall cardiovascular health.

#### *9.4.1.10 Durability and sweat resistance*

Materials and coatings in both the fitness tracker and patch antenna are chosen for durability and resistance to moisture. This ensures functionality even during intense workouts and exposure to sweat.

#### *9.4.1.11 Security and privacy*

Patch antennas contribute to secure wireless data transmission. Privacy features within fitness trackers protect sensitive health data, including heart rate information.

The integration of patch antennas in wearable fitness trackers, particularly for real-time heart rate monitoring, exemplifies the transformative impact of wireless communication technology on personal health management. This application not only empowers users with immediate insights but also contributes to a comprehensive approach to fitness, combining real-time monitoring with long-term trend analysis and personalized exercise recommendations. The effective synergy between patch antennas and heart rate monitoring represents a paradigm shift in health and wellness through wearable technology.

## 9.5 MICROSTRIP ANTENNAS IN BIOMEDICAL APPLICATIONS

Microstrip antennas find applications in various fields, including biomedical applications. The design characteristics of microstrip antennas make them suitable for integration into biomedical devices for wireless communication, sensing, and monitoring. Here are some aspects of microstrip antennas in biomedical applications:

### 9.5.1 Microstrip antennas in electrocardiogram (ECG) monitoring

ECG monitoring is a crucial aspect of cardiovascular health assessment, providing insights into the heart's electrical activity. The integration of microstrip antennas enhances the capabilities of ECG monitoring systems, contributing to real-time data transmission and wireless connectivity.

#### *9.5.1.1 Compact design for ECG sensors*

Microstrip antennas are selected for ECG monitoring due to their compact and low-profile design. This is particularly advantageous for the integration of antennas into small ECG sensors, ensuring unobtrusive and comfortable wear for users.

#### *9.5.1.2 Wireless connectivity in ECG systems*

Microstrip antennas play a pivotal role in establishing wireless communication between ECG sensors and external devices. This wireless connectivity eliminates the need for cumbersome wires, providing greater mobility to users during ECG monitoring.

#### *9.5.1.3 Real-time ECG data transmission*

The efficient design of microstrip antennas facilitates real-time transmission of ECG data from the sensor to external receivers or monitoring devices. This capability is essential for immediate monitoring and timely detection of cardiac irregularities.

#### *9.5.1.4 Conformal design for skin contact*

Flexible microstrip antennas support the conformal design of ECG sensors, ensuring that the sensors maintain consistent contact with the skin. This enhances the accuracy of ECG readings and provides a comfortable experience for users.

#### *9.5.1.5 Biocompatible materials*

Materials used in microstrip antennas can be selected for biocompatibility, ensuring they are safe for prolonged skin contact. This is critical for ECG monitoring, where sensors are in direct contact with the user's body.

#### *9.5.1.6 Low power consumption for prolonged monitoring*

Microstrip antennas are designed for low power consumption, optimizing the energy efficiency of ECG sensors. This is essential for prolonged monitoring periods, allowing users to wear the sensors for extended durations without frequent battery replacements.

#### *9.5.1.7 Enhanced mobility in ambulatory ECG monitoring*

Microstrip antennas contribute to ambulatory ECG monitoring by providing wireless connectivity. Users can go about their daily activities without the restriction of wired connections, promoting a more natural and unintrusive monitoring experience.

#### *9.5.1.8 Secure wireless communication*

Microstrip antennas play a crucial role in ensuring secure wireless communication in ECG monitoring systems. This addresses privacy concerns associated with transmitting sensitive health data, ensuring confidentiality and data integrity.

#### *9.5.1.9 Remote ECG monitoring and telemedicine*

The integration of microstrip antennas enables remote ECG monitoring, allowing healthcare professionals to receive real-time ECG data. This capability is particularly valuable for telemedicine applications, facilitating remote healthcare consultations and interventions.

Microstrip antennas significantly enhance the functionality of ECG monitoring systems, providing a wireless and compact solution for real-time data transmission. The seamless integration of these antennas into ECG sensors contributes to the advancement of cardiovascular health monitoring, offering improved mobility, comfort, and accessibility for users.

## 9.6 DIPOLE ANTENNAS IN BIOMEDICAL APPLICATIONS

Dipole antennas, known for their simple yet effective design, find diverse applications in the field of biomedical engineering. Their characteristics make them suitable for various wireless communication and sensing applications within the healthcare domain.

### 9.6.1 Dipole antennas in MRI (magnetic resonance imaging)

Magnetic Resonance Imaging (MRI) is a powerful medical imaging technique that utilizes strong magnetic fields and radiofrequency (RF) signals to produce detailed images of internal structures. Dipole antennas are crucial components in MRI applications, allowing wireless communication within the MRI environment.

#### 9.6.1.1 MRI-compatible antennas

In the MRI setting, conventional electronic devices can cause interference due to the intense magnetic fields and RF signals. Dipole antennas are specifically designed to be MRI-compatible, addressing the challenges associated with the unique environment of the MRI machine.

#### 9.6.1.2 Wireless communication in MRI

Dipole antennas play a pivotal role in enabling wireless communication within the MRI. They are integrated into medical implants or devices that require data transmission, reception of commands, or real-time monitoring during MRI procedures, eliminating the need for physical connectors.

#### 9.6.1.3 Implantable medical devices

A prominent application of dipole antennas in MRI is within implantable medical devices. Examples include pacemakers, neurostimulators, or drug delivery systems, where these antennas facilitate communication between the implanted device and external equipment during MRI scans.

#### 9.6.1.4 Signal transmission and reception

Dipole antennas in MRI serve a dual purpose – they transmit RF pulses to excite protons in the body, and they receive the resulting signals for image reconstruction. This functionality is essential for the generation of high-quality MRI images.

#### 9.6.1.5 Biocompatibility

Materials used in dipole antennas for MRI applications must be biocompatible to ensure patient safety. These antennas are often encapsulated or coated with materials that are compatible with the human body.

#### 9.6.1.6 MRI safety and image quality

Dipole antennas are carefully designed to minimize any adverse effects on MRI image quality. Their presence should not introduce artifacts or distortions that could compromise the diagnostic accuracy of the imaging process.

#### 9.6.1.7 Flexibility and conformal design

Flexibility is an advantageous feature of dipole antennas in MRI applications. They are designed to be flexible to conform to the shape of the implant or device, ensuring patient comfort and a secure fit during MRI procedures.

#### 9.6.1.8 *Frequency considerations*

Dipole antennas in MRI applications are optimized for the specific frequency range used in MRI RF pulses. This optimization ensures the efficient transmission and reception of signals within the MRI environment, contributing to the overall effectiveness of the imaging process.

#### 9.6.1.9 *Low power consumption*

Efficient power usage is a key characteristic of dipole antennas in MRI. They are designed for low power consumption to operate effectively during MRI procedures without causing excessive energy deposition or heating.

#### 9.6.1.10 *Research advancements*

Ongoing research focuses on advancing the performance of dipole antennas in MRI applications. This includes addressing challenges related to signal strength, minimizing artifacts, and ensuring compatibility with evolving MRI technologies.

Dipole antennas are integral components in MRI applications, contributing to the seamless integration of wireless communication within the MRI environment. Their design considerations for biocompatibility, flexibility, and optimization for specific frequencies enhance patient safety and diagnostic capabilities in the realm of MRI.

## 9.7 HELICAL ANTENNAS IN BIOMEDICAL APPLICATIONS

Helical antennas, with their unique spiral structure, find versatile applications in the field of biomedical engineering. These antennas offer specific advantages that make them suitable for various wireless communication and sensing applications within healthcare.

### 9.7.1 Overview of biotelemetry systems

**Definition:** Biotelemetry involves the remote transmission of biological data, facilitating continuous monitoring without physical connections.

**Purpose:** Enhancing patient mobility and comfort while ensuring real-time data collection for healthcare and research.

### 9.7.2 Components of a biotelemetry system

#### 9.7.2.1 *Biomedical sensors*

Biomedical sensors capture vital signs and physiological parameters. Helical antennas enable wireless data transmission from these sensors to external receivers.

#### 9.7.2.2 *Data processing unit*

The processing unit interprets transmitted data and performs initial processing. Helical antennas connect the processing unit to biomedical sensors, facilitating data flow.

#### 9.7.2.3 *External receiver*

External receivers collect transmitted data and relay it to central monitoring stations. Helical antennas on the external receiver ensure the reception of data from multiple biomedical sensors.

### 9.7.3 Role of helical antennas in biotelemetry systems

#### 9.7.3.1 *Wireless data transmission*

Helical antennas facilitate the wireless transmission of physiological data. Enhances patient mobility, eliminates the need for physical connections, and ensures seamless data flow.

#### 9.7.3.2 *Range and coverage*

Helical antennas contribute to extending communication range and coverage. Enables reliable data transmission over distances, crucial for monitoring patients in diverse environments.

### 9.7.4 Design considerations for helical antennas in biotelemetry

#### 9.7.4.1 *Frequency optimization*

Helical antennas are designed for specific frequencies optimized for biotelemetry applications. Proper frequency selection ensures efficient data transmission and reception, minimizing interference.

#### 9.7.4.2 *Compact form factor*

Helical antennas are designed to be compact, ensuring seamless integration with small biomedical sensors. Enhances aesthetics and minimizes impact on user comfort, especially in wearable devices.

#### 9.7.4.3 *Biocompatibility*

Materials used in helical antennas must be biocompatible for long-term contact with the human body. Biocompatible materials reduce the risk of skin irritation or adverse reactions, ensuring patient comfort.

### 9.7.5 Applications of helical antennas in biotelemetry

#### *9.7.5.1 Continuous patient monitoring*

Helical antennas enable continuous monitoring of patients in hospitals or home settings. Healthcare providers receive real-time data, allowing for timely interventions and personalized care.

#### *9.7.5.2 Research and clinical trials*

Biotelemetry systems with helical antennas contribute to research and clinical trials, providing researchers with continuous, wireless data streams. This facilitates the collection of high-quality data for scientific analysis and medical advancements.

### 9.7.6 Challenges and future developments

#### *9.7.6.1 Power consumption*

Helical antennas must operate efficiently to minimize power consumption, especially in battery-powered biomedical sensors. Ongoing research focuses on optimizing antenna designs for lower power requirements.

#### *9.7.6.2 Interference mitigation*

Biotelemetry systems may face interference in crowded environments. Advanced signal processing and frequency management techniques are explored to mitigate interference challenges.

## 9.8 ISM BAND ANTENNAS IN MEDICAL APPLICATIONS

ISM (Industrial, Scientific, and Medical) band antennas are designed to operate within the unlicensed frequency bands allocated for non-specific purposes. In medical applications, ISM band antennas play a crucial role in enabling wireless communication, data transmission, and connectivity for various healthcare devices and systems. Common ISM bands include 433 MHz, 868 MHz, and 2.4 GHz, providing options for different applications.

## 9.9 COMPONENTS OF ISM BAND ANTENNAS IN MEDICAL DEVICES

### 9.9.1 Wireless medical devices

Medical devices, such as sensors, monitors, and implants, often incorporate wireless communication capabilities. ISM band antennas facilitate wireless data transmission between these devices and external receivers.

### 9.9.2 External receivers and gateways

External receivers or gateways collect data from wireless medical devices. ISM band antennas on these external components ensure reliable reception of data transmitted by medical devices.

## 9.10 ROLE OF ISM BAND ANTENNAS IN MEDICAL APPLICATIONS

### 9.10.1 Wireless connectivity

ISM band antennas enable wireless connectivity, allowing medical devices to communicate without physical connections. This enhances mobility, reduces cable clutter, and supports the development of wearable and implantable medical devices.

### 9.10.2 Data transmission

ISM band antennas facilitate the transmission of medical data, including vital signs, patient monitoring information, or diagnostic data. This ensures real-time data transfer for timely decision-making by healthcare professionals.

## 9.11 APPLICATION OF ISM BAND ANTENNAS IN WEARABLE HEALTH DEVICES

The use of ISM band antennas enhances the functionality of wearables, providing real-time health data to users and healthcare professionals.

### 9.11.1 ISM bands for wearables

ISM bands, particularly 2.4GHz, are commonly used in wearables for their unlicensed status and compatibility with small form factor devices. ISM band antennas enable wearables to establish wireless connections with smartphones, tablets, or central monitoring systems.

### 9.11.2 Components and design considerations

#### *9.11.2.1 Biometric sensors*

Wearable health devices incorporate sensors for monitoring parameters like heart rate, activity levels, or temperature. ISM band antennas facilitate wireless communication, allowing these sensors to transmit data to the device's processing unit.

#### 9.11.2.2 *Processing unit*

The processing unit in wearables analyses sensor data and manages communication with external devices. ISM band antennas connect the processing unit to external devices, supporting the transmission of processed health data.

#### 9.11.2.3 *Form factor and size*

ISM band antennas in wearables are designed to be compact and fit within the limited space of small devices. Allows for the integration of antennas without compromising the aesthetics or comfort of the wearable.

#### 9.11.2.4 *Applications in wearable health devices*

ISM band antennas enable wearables to continuously monitor health parameters, such as heart rate or blood pressure. Users receive real-time data on their health status, promoting proactive health management.

#### 9.11.2.5 *Fitness tracking*

Wearables equipped with ISM band antennas track fitness metrics, including steps taken, distance covered, and calories burned. Wireless transmission of fitness data to smartphones or fitness apps enhances user engagement and goal tracking.

### 9.11.3 Challenges and future developments

#### 9.11.3.1 *Power efficiency*

Wearable devices operate on limited battery power. Ongoing research focuses on optimizing ISM band antenna designs for lower power consumption, extending wearable device battery life.

#### 9.11.3.2 *Interoperability and standards*

Ensuring interoperability between different wearables and external devices. Standardization efforts aim to enhance the compatibility and seamless integration of wearables across various platforms.

## 9.12 SLOT ANTENNAS

Slot antennas, known for their low-profile design and efficient radiation characteristics, find diverse applications in the medical field. Their unique properties make them suitable for various medical devices and systems that require wireless communication, sensing, and imaging. Slot antennas are formed by creating openings or slots in a conductive surface, typically a metal plate. Slot antennas are integrated into medical imaging devices such as MRI machines. In devices like pacemakers or neurostimulators, slot antennas contribute to wireless communication. Slot antennas enable wireless data transmission in medical devices, reducing the need for physical connectors. Slot antennas are designed for specific frequency bands based on the application requirements.

## 9.13 WIRELESS CAPSULE ENDOSCOPY: UTILIZING SLOT ANTENNAS FOR DIAGNOSTIC IMAGING

WCE is a cutting-edge medical application that leverages slot antennas to revolutionize the diagnosis and visualization of the gastrointestinal (GI) tract. This innovative technology employs a small capsule equipped with a camera and sensors, offering a non-invasive approach to capturing high-resolution images and diagnostic data throughout the digestive system. Slot antennas integrated into the capsule play a critical role in enabling wireless communication, allowing real-time transmission of crucial medical information to external receivers.

### 9.13.1 Overview of wireless capsule endoscopy

WCE serves the purpose of providing comprehensive visualization and diagnosis of GI disorders, including bleeding, tumors, or inflammatory conditions. The key advantage lies in its non-invasive nature, offering patients a more comfortable alternative to traditional endoscopic procedures. The capsule's ability to navigate the entire digestive tract provides detailed imaging that is often inaccessible through conventional methods.

### 9.13.2 Components and role of slot antennas

Slot antennas are strategically integrated into the capsule, working in tandem with a small camera and sensors responsible for capturing detailed images and diagnostic data. These antennas play a crucial role in establishing wireless communication, enabling the continuous transmission of data from the capsule to an external receiver worn by the patient.

### 9.13.3 Wireless communication in WCE

The wireless communication facilitated by slot antennas allows for seamless data transmission from the capsule throughout its journey in the GI tract. Patients wear a compact external receiver unit that collects and stores the transmitted data. Slot antennas ensure a reliable wireless link, enabling real-time data transfer, and allowing patients the freedom to move during the examination.

### 9.13.4 Design considerations for slot antennas in WCE

Slot antennas in WCE are designed with specific frequency bands in mind, optimizing them for efficient data transmission and compliance with medical regulations. The compact form factor of these antennas is crucial to fit within the limited space of the capsule, ensuring their integration does not compromise the overall size and functionality of the diagnostic device.

### 9.13.5 Advantages and challenges

The advantages of WCE, empowered by slot antennas, are evident in its non-invasiveness and ability to provide comprehensive imaging. However, challenges include the limited wireless range of the capsule, necessitating close proximity to the external receiver, and potential interference in environments with multiple wireless signals.

### 9.13.6 Future developments

Ongoing research focuses on further miniaturizing slot antennas, aiming to create even smaller capsules for enhanced patient comfort. Additionally, improving the biocompatibility of materials used in slot antennas remains a priority to ensure the safety of patients during extended use.

In conclusion, the integration of slot antennas in WCE represents a groundbreaking advancement in medical diagnostics. This technology not only offers a patient-friendly alternative to traditional endoscopy but also provides healthcare professionals with comprehensive and real-time imaging of the GI tract. As technology continues to evolve, ongoing research will address challenges and contribute to the continual improvement of WCE, making it an increasingly valuable tool in the field of medical diagnostics.

## 9.14 WIRELESS SENSOR ANTENNAS IN MEDICAL APPLICATIONS

Wireless sensor antennas have become indispensable components in various medical applications, contributing to the evolution of patient monitoring, diagnostics, and treatment. These antennas enable the seamless integration of sensors into medical devices, fostering wireless communication and data transmission critical for real-time healthcare monitoring. Wireless sensor antennas enable the remote monitoring of patients' vital signs and physiological parameters.

### 9.14.1 Role of wireless sensor antennas in patient monitoring

Patient monitoring is a critical aspect of healthcare that involves continuous observation of a patient's vital signs and physiological parameters.

The integration of wireless sensor antennas in patient monitoring systems has revolutionized healthcare by enabling real-time data collection, remote monitoring, and timely interventions. Here, we delve into the detailed role of wireless sensor antennas in patient monitoring:

#### 9.14.1.1 Continuous monitoring

Wireless sensor antennas facilitate continuous monitoring of a patient's vital signs, including heart rate, blood pressure, respiratory rate, and temperature. This enables healthcare providers to access real-time data without the need for constant physical presence, allowing for more proactive patient care.

#### 9.14.1.2 Remote patient monitoring

Wireless sensor antennas enable the transmission of patient data to remote monitoring stations or healthcare facilities. Patients can be monitored from a distance, making it especially valuable for home healthcare, chronic disease management, or post-surgery recovery.

#### 9.14.1.3 Emergency alerts

Wireless sensor antennas are crucial for transmitting emergency alerts in case of critical events or abnormal readings. Enables rapid response from healthcare providers, allowing them to intervene promptly in emergency situations.

#### 9.14.1.4 Biometric data transmission

Wireless sensor antennas facilitate the wireless transmission of biometric data captured by medical sensors. This includes ECG data, blood glucose levels, oxygen saturation, and other relevant physiological parameters. Provides a comprehensive overview of the patient's health status, aiding in the early detection of abnormalities.

#### 9.14.1.5 Wearable health devices

Wireless sensor antennas are integrated into wearable health devices, such as fitness trackers and smartwatches. Allows individuals to continuously monitor their health parameters, and the data is wirelessly transmitted to smartphones or central monitoring platforms. Supports proactive health management and encourages users to make informed lifestyle choices.

#### 9.14.1.6 Data transmission to central monitoring stations

Wireless sensor antennas establish a communication link between wearable sensors or implanted devices and central monitoring stations.

Centralized data collection allows healthcare providers to track trends, analyze historical data, and make informed decisions regarding patient care plans.

#### 9.14.1.7 *Bi-directional communication in implantable devices*

In implantable medical devices like pacemakers or insulin pumps, wireless sensor antennas enable bi-directional communication. This allows for the wireless transmission of patient-specific data to external devices and, in some cases, facilitates remote adjustments or programming of the implantable device.

#### 9.14.1.8 *Real-time intervention*

Real-time data transmission through wireless sensor antennas enables healthcare providers to intervene promptly in case of deteriorating health conditions. Timely interventions can prevent complications, reduce hospitalization rates, and improve overall patient outcomes.

#### 9.14.1.9 *User-friendly monitoring*

Wireless sensor antennas contribute to user-friendly patient monitoring by eliminating the need for cumbersome wired connections. This enhances patient comfort and mobility, allowing individuals to move freely while being monitored.

#### 9.14.1.10 *Data security and privacy*

Wireless sensor antennas are designed with security protocols to ensure the confidentiality and privacy of patient data during transmission. Safeguards sensitive medical information, addressing concerns related to data security and patient privacy.

#### 9.14.1.11 *Integration with electronic health records (EHR)*

Wireless sensor data can be integrated seamlessly with electronic health records for a comprehensive view of the patient's medical history. Provides healthcare professionals with a holistic understanding of the patient's health, supporting better-informed decisions.

## 9.15 FUTURE SCOPE OF BIOMEDICAL ANTENNAS IN HEALTHCARE

The advancements in biomedical antennas have set the stage for a transformative future in healthcare technology. As we navigate the intricate landscape of wireless connectivity in medical applications, several promising avenues and future directions emerge, shaping the trajectory of this dynamic field:

### 9.15.1 Miniaturization and integration

Ongoing research will focus on further miniaturizing biomedical antennas, enabling their seamless integration into smaller and more discreet medical devices. This will enhance patient comfort, promote the development of minimally invasive devices, and allow for unobtrusive health monitoring.

### 9.15.2 5G Technology integration

As 5G technology becomes more widespread, the integration of biomedical antennas with 5G networks holds immense potential for faster and more reliable wireless communication in healthcare. Enhanced data transfer speeds and reduced latency will enable real-time monitoring, telesurgery, and the development of more sophisticated medical applications.

### 9.15.3 Advanced materials and biocompatibility

Continued exploration of advanced materials and coatings for biomedical antennas to improve biocompatibility and reduce the risk of adverse reactions.

### 9.15.4 Energy harvesting technologies

Integration of energy harvesting technologies with biomedical antennas to power devices using ambient energy sources. Prolonged battery life, reduced dependence on external power sources, and increased sustainability for implantable and wearable medical devices.

### 9.15.5 AI and data analytics integration

Integration of artificial intelligence (AI) and advanced data analytics with biomedical antennas for real-time interpretation of health data.

### 9.15.6 Internet of medical things (IoMT)

The proliferation of IoMT, where interconnected medical devices communicate seamlessly, will drive the evolution of biomedical antennas.

### 9.15.7 Standardization and interoperability

Continued efforts in standardizing communication protocols and ensuring interoperability among various medical devices. Seamless integration of diverse biomedical antennas into unified healthcare systems, fostering a more cohesive and efficient healthcare infrastructure.

### 9.15.8 Wearable health technology innovations

Advancements in wearable health technology, with more sophisticated sensor arrays and antennas, will redefine personal health monitoring. These improvements will lead to increased accuracy, expanded functionality, and greater user engagement in health and wellness tracking.

### 9.15.9 Telemedicine and remote surgery

Biomedical antennas will play a pivotal role in enabling advanced telemedicine and remote surgery applications. Improved connectivity will facilitate remote consultations, surgeries, and medical procedures, expanding access to healthcare services globally.

### 9.15.10 Cybersecurity and privacy measures

Implementation of robust cybersecurity measures and privacy protocols for protecting sensitive health data transmitted through biomedical antennas. Enhanced patient trust, regulatory compliance, and safeguarding against potential security breaches.

The exploration of advancements in biomedical antennas within this book chapter illuminates the dynamic landscape of wireless connectivity in medical applications. The diverse designs and innovative approaches discussed herein underscore the pivotal role antennas play in the evolution of modern healthcare technology. This comprehensive journey through various antenna types and their applications in medical scenarios provides valuable insights into the cutting-edge developments shaping the future of patient care.

As we gaze into the future, the convergence of these developments will not only refine the capabilities of biomedical antennas but also revolutionize the landscape of healthcare delivery. From personalized medicine and real-time diagnostics to the seamless integration of medical devices into interconnected health ecosystems, the future scope of biomedical antennas holds the promise of a more accessible, efficient, and patient-centric healthcare paradigm. Researchers, engineers, and healthcare professionals alike stand at the forefront of this transformative journey, poised to shape the future of healthcare through the continued evolution of biomedical antennas.

## 9.16 CONCLUSION

In conclusion, this book chapter serves as a comprehensive guide for researchers, engineers, and healthcare professionals navigating the intricate world of biomedical antennas. The amalgamation of diverse antenna designs and their applications in medical contexts paints a vivid picture of the transformative impact these technologies have on healthcare delivery. As we stand at the intersection of innovation and healthcare, the advancements explored in this chapter not only reflect the state of the art but also pave the way for a future where wireless connectivity seamlessly integrates into the fabric of medical practices, ultimately improving patient outcomes and shaping the landscape of healthcare for years to come.

## REFERENCES


1. T. Zasowski, Ultra-Wideband Antennas for Biomedical Telemetry Applications, *IEEE Transactions on Antennas and Propagation,* vol. 57, no. 3, pp. 663–671, 2009.
2. J. M. Rabaey et al., Health Gear: A Real-Time Wearable System for Monitoring and Analysing Physiological Signals, *IEEE Transactions on Biomedical Circuits and Systems,* vol. 4, no. 3, pp. 179–188, 2010.
3. M. Islam, The Internet of Things for Health Care: A Comprehensive Survey, *IEEE Access,* vol. 3, pp. 678–708, 2015.
4. M. O. F. Howlader and T. P. Sattar, Miniaturization of Dipole Antenna for Low Frequency Ground Penetrating Radar, *Progress in Electromagnetics Research C,* vol. 61, pp. 161–170, 2016.
5. N. Sharma, Wearable Health Monitoring Systems: A Review, *Sensors & Transducers Journal,* vol. 225, no. 3, pp. 67–73, 2019.
6. A. M. A. Hafiz et al., A Review on Wearable Photoplethysmography Sensors and Their Potential Future Applications in Health Monitoring, *Journal of Personalized Medicine,* vol. 9, no. 3, p. 20, 2019.
7. N. Sharma, Wearable Health Monitoring Systems: A Review, *Sensors & Transducers Journal,* vol. 225, no. 3, pp. 67–73, 2019.
8. J. A. Zhang, 5G Technologies for Healthcare: Challenges and Opportunities, *Journal of King Saud University - Computer and Information Sciences,* vol. 34, no. 9, pp. 6949–6976, 2022.
9. S. Abadal, Wireless Communications for Healthcare: A Survey, *Wireless Networks,* vol. 26, no. 3, pp. 1665–1681, 2020.
10. M. U. Hadi, and G. Murtaza, Fibre Wireless Distributed Antenna Systems for 5G and 6G Services, *Electronics,* vol. 12, pp. 1–16, 2022.

Chapter 10

# Analysis of antenna for biomedical applications

*R. Prabha, V. Subashini, M. Aishwarya, B. Hemalatha, and A. Sadhana*

## 10.1 INTRODUCTION

Antennas are extensively employed as an important element in wireless communications systems. Fractal UWB planar antenna for biomedical applications, especially for microwave and mm-wave medical imaging [1]. The antenna has a quasi-omnidirectional radiation pattern which allows it to be placed on the external body and it occupies a compact area of $13 \times 13\,mm^2$ which is flexible to incorporate with wearable garments or gadgets for different biomedical applications [2]. A hexagonal ring-shaped parasitic structure in the ground plane is designed at the frequency of 2.46 GHz. It is used to degree the frame temperature [3]. The biomedical patch antenna with a wireless data transfer system has been proposed. To provide efficient surface current distribution of the antenna, the slot pattern has been designed, which works at 20 GHz and works efficiently with small chip sizes and is easily incorporated with biomedical applications [4]. It is also used for various purposes such as mobile, satellites, and working at L and S bands. It would be helpful to destroy the virus tissues of tumours and hypothermia diseases by increasing the temperature [5]. Many research works have been undergone to detect breast cancer by microwave imaging, $18 \times 18\,mm^2$ monopole antenna, wearable prototype with 16 elements array antenna has been proposed [6]. An antenna is designed for an imaging system for S-band frequency with a 2 GHz. An antenna supplied by a coplanar waveguide has been proposed for in-body operations [7]. In Ref. [8], An antenna with a return loss of –24 dB resonates at 3.55 GHz V-shaped slotted antenna has been implemented in Ref. [9]. Hexagonal-shaped microstrip antenna for V2V communications [10], Vehicular to infrastructure, vehicle to vehicle communications [11]. A UWB Wearable Microstrip Patch Antenna Design for Wireless Applications on the Human Body. This chapter concludes that the antenna has high bandwidth efficiency, high security in communication, and the ability to penetrate the human body [12]. A flower bud-shaped flexible UWB antenna for health care applications is discussed. This study

DOI: 10.1201/9781003560487-10

illustrates that UWB antenna can be used as passive sensors in biomedical applications [13]. Trends in Artificial Neural Network-Based Microwave Device Design. This chapter explains the design of a wearable ultra-wideband antenna triangular patch for health application [14]. Ultra wideband for biomedical application UWB antenna has high security in communication, and has the ability to penetrate the human body [15]. In Ref. [16], wide applications of antennas has been overviewed and play a major and important role in communication systems [17]. U-shaped patch antenna for transmitter and receiver side suitable for sub 6 GHz range is discussed. And here, the rectangular patch antenna has been proposed for biomedical operations [18] honeycomb structured reconfigurable antenna for sub 6 GHz frequency range is analysed.

## 10.2 DESIGN METHODOLOGY

### 10.2.1 Development of rectangular patch antenna

The methodology involves various steps to design the antenna for biomedical applications. It is easy and compact to design and manufacture into various devices. The rectangular shape ensures optimal performance in specific communication bands. These specifications ensure cost-effectiveness and pave the way for many applications. Its low-profile design is well suited for medical applications such as wearable devices and implantable devices. Firstly, factors like frequency range, size constraints and type of antenna should be considered. Secondly, The CST tool is a simulation software used to optimize the dimensions and geometry to obtain the desired results. Thirdly, the antenna should be fabricated based on the optimized design. At last, this fabricated antenna should be integrated with a biomedical system. We evaluate the parameters such as return loss, radiation pattern, and VSWR to ensure that it meets the performance criteria. After optimization the antenna is fabricated. This antenna uses copper (lossy) for the ground and radiating plane and FR4 for the substrate. Then the antenna undergoes a testing process in a controlled environment to validate its performance.

## 10.3 PROPOSED WORK

The antenna is designed and analysed using CST Studio Suite. The results such as frequency range and flexibility can be analysed using the CST tool. The antenna is designed with the size of 40×40×1.6 mm using an FR-4 substrate material having the dielectric constant ($\varepsilon_r$) of 4.4. The operating frequency range of the antenna is 2.4 GHz. The S parameter, VSWR, and radiation pattern have been analysed. Then our antenna is planned to be integrated to make it a promising candidate for various biomedical applications. Rectangular patch antennas have many advantages when applied in biomedical applications. These antennas exhibit good impedance-matching properties and ensure efficient energy transfer between the antenna and associated electronic circuits. This facilitates widespread adoption in various biomedical applications, such as implantable devices, wearable health monitors, and medical imaging systems. Here, the copper is used as a patch. Copper which is in annealed form has high significance in the realm of antennas in medical fields. The substrate is made of FR4. Additionally, it is frequently as a dielectric material because of its electrical characteristics, which make it appropriate for antenna design. And the ground is made of the same annealed copper. The flow chart is shown in Figure 10.1. The rectangular patch antenna is shown in Figure 10.2. The performance analysis of the proposed antenna is shown in Table 10.1. The dimension of the proposed antenna is shown in Table 10.2.

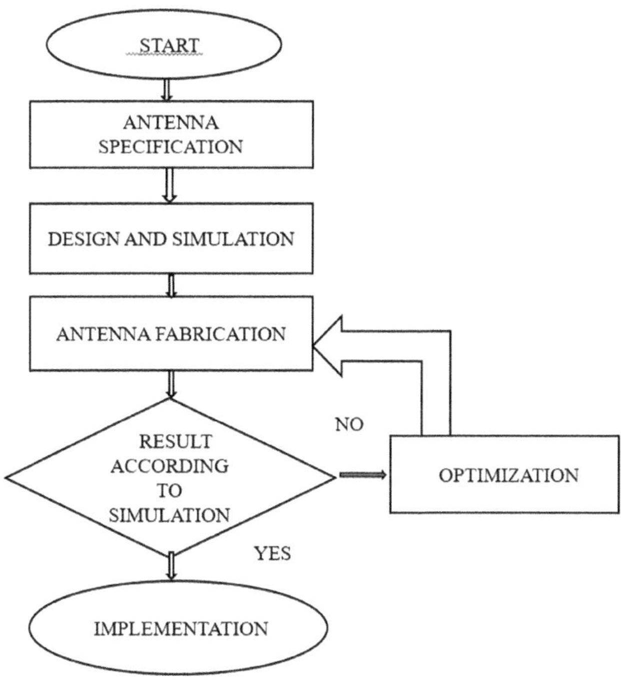


*Figure 10.1* Flowchart.

*Figure 10.2* Rectangular patch antenna.

*Table 10.1* Performance analysis of the proposed antenna

| *Frequency* | *Return loss* | *VSWR* | *Bandwidth* |
|---|---|---|---|
| 2.4 GHz | −14 dB | 1.2 | 2.425–2.353 GHz |

*Table 10.2* Dimension of the proposed antenna

| | | |
|---|---|---|
| Width of substrate | $W$ | 65 |
| Length of substrate | $L$ | 65 |
| Height of substrate | $H$ | 1.6 |
| Width of patch | $W_p$ | 36 |
| Length of patch | $L_p$ | 28.68 |
| Height of patch | $G$ | 0.035 |
| Width of feedline | $W_f$ | 1.44 |

## 10.4 RESULTS AND DISCUSSION

The obtained output parameters are return loss, VSWR, radiation pattern, and efficiency. The operating resonant frequency is 2.4 GHz, as shown in Figure 10.3.

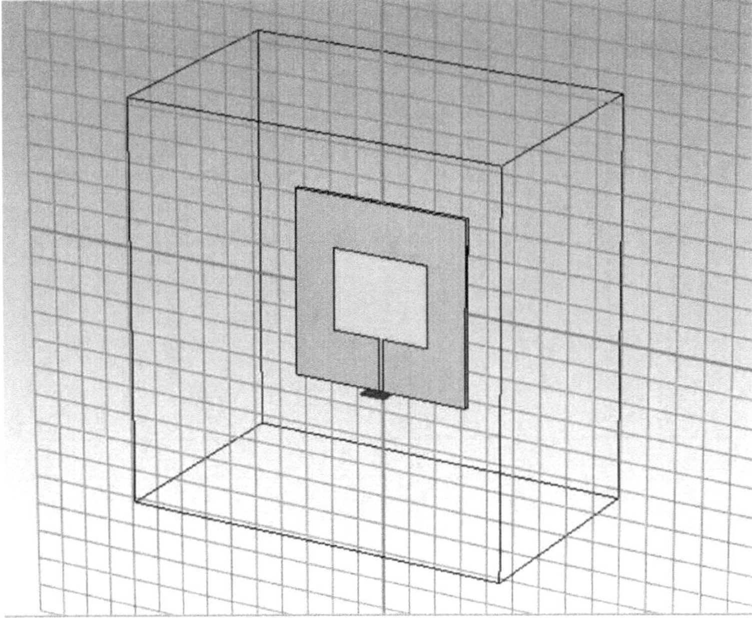

*Figure 10.3* Proposed antenna design.

### 10.4.1 Return loss

The ability of an antenna to send and receive signals is measured by its return loss. It is expressed as decibels. Higher return loss shows that the antenna has better performance. Return loss is an important parameter for measuring the efficiency and performance of antennas. It quantifies the degree of impedance matching, which is vital for optimizing signal transmission and minimizing power loss and reflections in RF systems, as shown in Figure 10.4.

### 10.4.2 Voltage standing wave ratio

The voltage standing wave ratio functions as a measure of radiofrequency power transmission efficiency. From the below graph it can be seen that the VSWR has the dip at the 2.4 GHz for the frequency in range. It represents a perfect impedance match between the antenna and the connected devices. VSWR values represent poorer impedance matching and increased power reflections, which can lead to signal loss and degradation in RF systems. VSWR is an important parameter for ensuring efficient signal transmission and minimizing signal loss in antennas and RF devices, as shown in Figure 10.5.

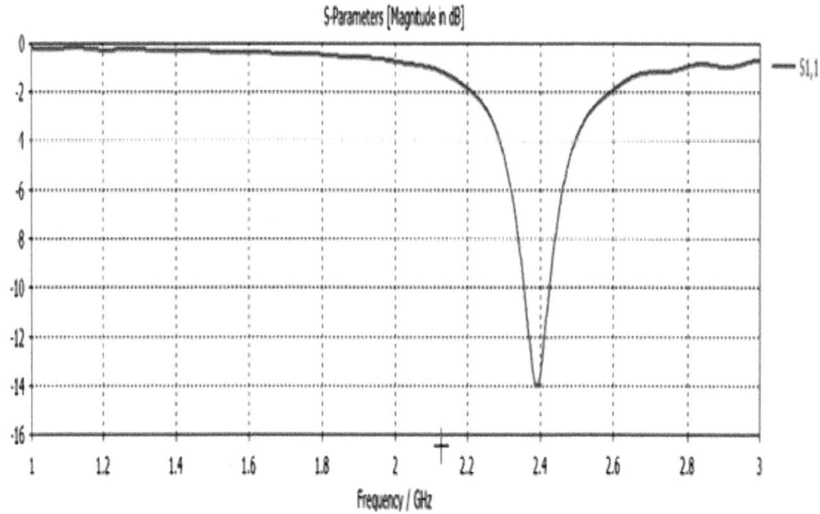


*Figure 10.4* Return loss.

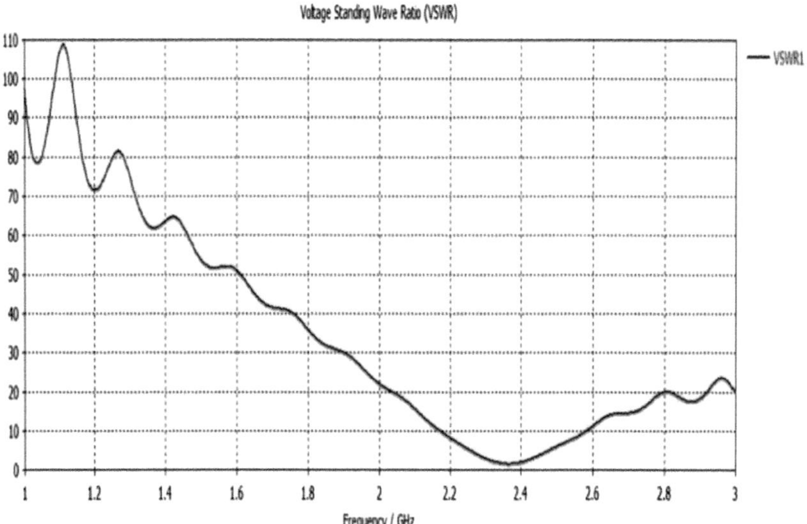


*Figure 10.5* VSWR.

### 10.4.3 Radiation pattern

The directional characteristics of an antenna or other electromagnetic source are represented graphically by the radiation pattern. It shows how the electromagnetic energy is distributed in different directions. It also

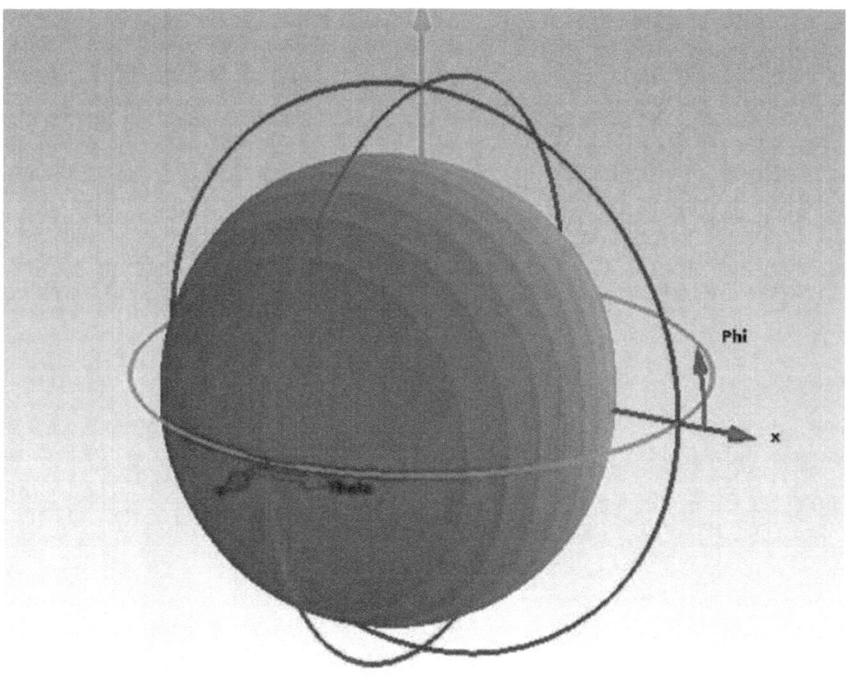


*Figure 10.6* Radiation pattern.

provides valuable information about the directional characteristics of the antenna, showing how the power is radiated and varies with respect to other spatial angles, as shown in Figure 10.6.

### 10.4.4 Surface current distribution

A graphic representation of an antenna's two-dimensional power distribution shows how power propagates simultaneously within and around the antenna. This information is often used to provide a visual depiction of an antenna's radiation pattern or reception pattern, as shown in Figure 10.7.

### 10.4.5 Bandwidth

The range of frequencies that an antenna can properly function over is referred to as its bandwidth.

The analyser measures and displays the amplitude of RF signals over a given frequency range, as shown in Figure 10.8. They are used to determine the frequency response and bandwidth of antennas, identify interference

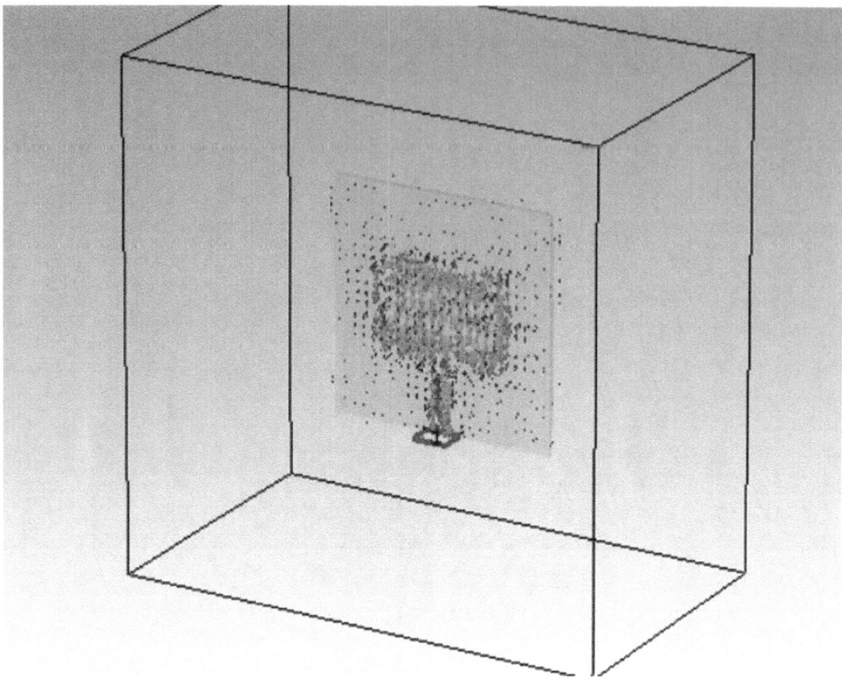

*Figure 10.7* 2D view of surface current distribution.

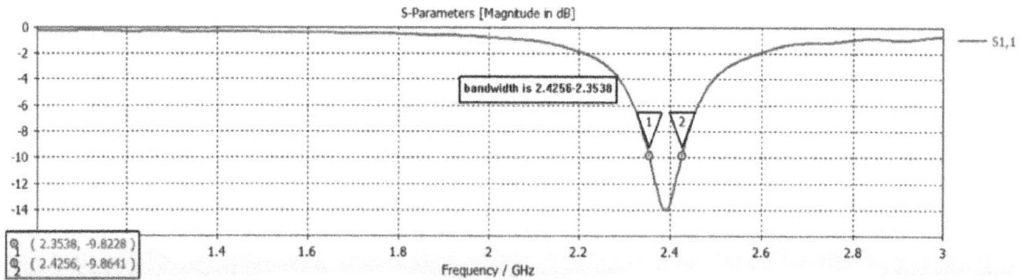


*Figure 10.8* Bandwidth calculation.

and assess the performance of RF systems, as shown in Figure 10.9. VNAs are sophisticated RF test instruments which measure the amplitude and phase of RF signals at various frequencies. They are widely used for antenna characterization, measurement of return loss and VSWR, gain, and radiation patterns, as shown in Figure 10.10.

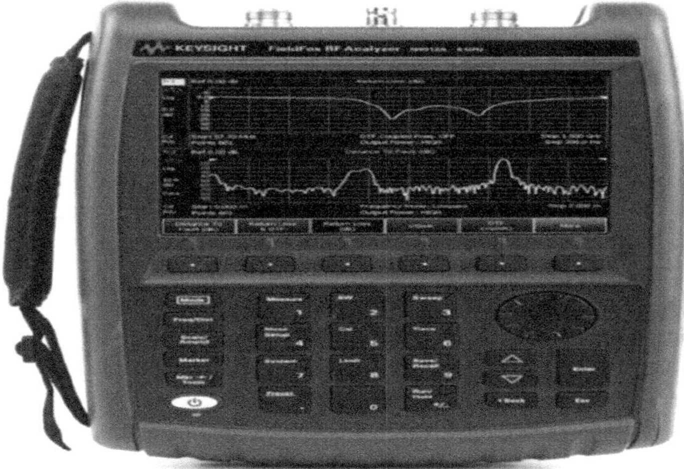


*Figure 10.9* Field flux RF analyser top view.

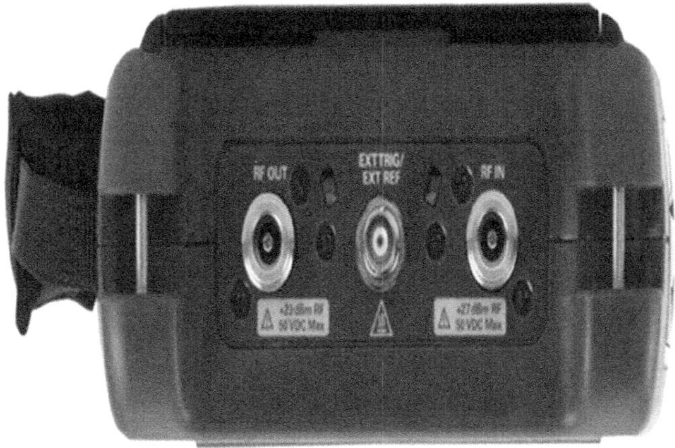


*Figure 10.10* Field flux RF analyser side view.

## 10.5 CONCLUSION

A rectangular patch antenna was designed for biomedical applications. The antenna was designed to operate in the frequency 2.4 GHz. Parameters such as return loss, radiation pattern, and VSWR were analysed using CST. Biomedical applications need antennas that are specifically designed to operate in or around the human body. Using antennas in the biomedical field enables to transmit and receive data wirelessly. Wearable health monitoring devices utilize compact, low-profile antennas for real-time data transmission that enables healthcare and enables patients and medical practitioners to monitor physiological parameters. Additionally, antennas in RF ablation catheters and RFID tags for medication tracking are crucial for interventional procedures and drug management in the healthcare industry, contributing to improved patient care and safety. This technology facilitates real-time data transfer between medical devices.

## 10.6 FUTURE SCOPE

Although the antenna is configured to work in UWB same can be extended and made available to work in different frequency bands. A reconfigurable antenna can also be developed which switches itself according to the frequency of incoming signals. A wearable antenna can also be designed for the same application. Moreover, in the field of 5G communications, rectangular patch antennas hold potential for many applications. As 5G networks deploy higher frequencies and demand increased data rates, antenna technologies become important for efficient signal transmission. The adaptability and ease of fabrication of rectangular patch antennas make them candidates for use in 5G mobile devices, and other infrastructure components. Their development may contribute to enhancing the performance and reliability of 5G networks, supporting the vision of ultra-fast, low-latency wireless communications which enable transformative technologies such as augmented reality, virtual reality, and autonomous systems.

## REFERENCES


1. A. Iqbal, M.A.I Hasan, UWB planar antenna for biomedical medical devices, *IEEE Sensors Journal,* Vol. 23, 3232727, 2023.
2. Md.S. Islam, S.M. Kayser Aslam, A.K.M. Zakir Hossain, Low profile flexible planar monopole antenna for biomedical applications, *Engineering Science and Technology, an International Journal,* Vol. 35, 10, 2022.
3. K. Kumar, Design of flexible parasitic element patch antenna for biomedical applications, *Engineering Science and Technology, an International Journal,* Vol. 94, 143–153, 2020.
4. G.-S. Byum, A wireless data transfer by using a patch antenna for biomedical applications, *IEEE Sensors Journal,* Vol.10, 1270–1274, 2021.
5. R. Pratap, B. Singh, Miniaturization of microstrip patch antenna for biomedical applications, *International Journal of Electrical and Electronics Research,* Vol. 10, 1270–1274, 2021.

6. H. Bahramiabarghouei, Flexible 16 antenna array for microwave breast cancer detection, *IEEE Antenna Wiredless Propagation Letter*, Vol. 20, no. 7, pp. 1185–1189, Jul. 2021.
7. D. Fobar, L. Phillips, A. Wilhelm, P. Chapman, Considerations for training an artificial neural network for particle type identification, *IEEE Transactions on Nuclear Science*, Vol. 68, 2350–2357, 2021.
8. T. Lin, Y. Zhu, Beamforming design for large-scale antenna arrays using deep learning, *IEEE Wireless Communications Letters*, Vol. 9, 103–107, 2020.
9. R.A. Vatti, B. Arthi, P. Kumar, R. Prabha, Gain enhancement by utilizing hexagonal reflector based optimized textile antenna, *Journal of Ambient Intelligence and Humanized Computing*, Vol. 15, no. 18, pp. 258–266, 2021.
10. R. Lesnik, Gain enhancement of a compact implantable dipole for biomedical applications, *Engineering Science and Technology, an International Journal*, Vol. 35, 101112, 2022.
11. D. Kurup, A compact CPW fed UWB antenna with quad band notch characteristics for ISM band applications, *Progress in Electromagnetics Research*, Vol. 21, 3062415, 2021.
12. N. Ahmed Malil, T. Ajmal, P. Sant, M.U. Rehman, Compact size implantable antenna for biomedical applications, *IEEE Access*, Vol. 10, 9205355, 2022.
13. K. Niak, Flexible CPW-fed split-triangular shaped patch antenna for WIMAX applications, *IEEE Antennas Wireless Propagation Letters*, Vol. 20, no. 7, 1185–1189, Jul. 2021.
14 D. Fobar, L. Phillips, A. Wilhelm, P. Chaapman, Considerations for training an artificial neural network for particle type identification, *IEEE Transactions on Nuclear Science*, Vol. 68, 2350–2357, 2021.
15 T. Lin, Y. Zhu, Beamforming design for large-scale antenna arrays using deep learning, *IEEE Wireless Communication Letters*, Vol. 9, 103–107, 2020.
16. V. Kumar, On body measurements of SS-UWB patch antenna for WBAN applications. *International Journal of Electrical – Electronics Research*, Vol. 70, no 5, 668–675, May 2016.
17. B. Rajalakshmi, M. Jenath, V. Subashini, J. Alsamath, Sub 6 GHz Tx-Rx microstrip antenna for 5G applications. *IEEE*, 978-1-6654-6275-4, 2022.
18. V. Subashini, S. Chitra, A compact honeycomb-structured reconfigurable antenna with coplanar waveguide feed for multiband wireless applications, *Telecommunications and Radio Engineering*, Vol. 8, no. 2, 37–38, , 2024.

Chapter 11

# Optimized connectivity

## Antennas in industrial and medical applications for enhanced wireless communication

*M. Mary Adline Priya*

## 11.1 INTRODUCTION

The convergence of wireless communication technologies with industrial and medical applications has ushered in a new era of connectivity, offering unprecedented opportunities and challenges. This introduction sets the stage by outlining the growing importance of reliable and efficient wireless communication in both industrial and medical contexts. It highlights the unique requirements and constraints faced in these sectors, underscoring the need for antennas optimized to meet specific demands.

In an era defined by the relentless pace of technological evolution, the convergence of wireless communication technologies with industrial and medical applications has emerged as a transformative force. The amalgamation of these domains not only enhances operational efficiency in industries but also revolutionizes patient care and medical diagnostics. At the heart of this transformative wave lies the indispensable role of antennas, the unsung heroes enabling seamless and reliable wireless communication in the most demanding environments.

Wireless communication, once a convenience, is now a cornerstone in both industrial and medical sectors. In industries, the deployment of wireless systems has become synonymous with the pursuit of efficiency, automation, and real-time connectivity. Likewise, in medical applications, wireless communication facilitates the interconnected ecosystem of healthcare, enabling remote patient monitoring, implantable devices, and telemedicine. However, these advancements come with a unique set of challenges, and the optimization of antennas stands as the linchpin in overcoming these hurdles.

## 11.2 THE UNIQUE CHALLENGES

Industrial environments pose challenges that range from the cacophony of electromagnetic interference on factory floors to the necessity for communication resilience across vast operational landscapes. The reliability

 DOI: 10.1201/9781003560487-11

of wireless communication in these settings directly impacts operational efficiency, safety, and the overall efficacy of industrial processes. On the other hand, medical applications demand a level of precision and safety that transcends typical wireless communication constraints. Implantable medical devices, healthcare monitoring systems, and diagnostic equipment necessitate antennas that can operate seamlessly within the human body while adhering to stringent regulatory standards.

## 11.3 OVERVIEW OF ANTENNAS AND THEIR IMPORTANCE

Antennas are essential components in wireless communication systems, playing a crucial role in transmitting and receiving electromagnetic signals. They convert electrical signals into radio waves and vice versa, enabling the exchange of information between various devices and systems. Antennas are available in a wide range of shapes, sizes, and designs to cater to different applications, frequencies, and environments.

Antennas play a crucial role in wireless communication by converting electrical signals into radio waves for transmission and vice versa for reception. They are essential for devices like cell phones, IoT devices, and satellite systems to communicate effectively. Antennas are evolving to meet the increasing demands of the Internet of Things (IoT) and smart systems by offering greater directivity, flexibility in design, and improved signal quality. As technology advances, antennas are becoming more modularized to accommodate a variety of solutions and are crucial for the development of adaptive antennas for future wireless networks.

The importance of antennas in industrial and medical applications can be highlighted by their ability to facilitate efficient communication, data transfer, and control in various settings. In industrial applications, antennas play a vital role in enabling wireless communication between machines, sensors, and control systems, leading to improved productivity, safety, and cost-effectiveness. In medical applications, antennas are used in devices such as medical imaging systems, patient monitoring devices, and telemedicine systems, contributing to better patient care and outcomes.

The importance of antennas in industrial and medical applications can be highlighted by their ability to facilitate efficient communication, data transfer, and control in various settings. In industrial applications, antennas play a vital role in enabling wireless communication between machines, sensors, and control systems, leading to improved productivity, safety, and cost-effectiveness. In medical applications, antennas are used in devices such as medical imaging systems, patient monitoring devices, and telemedicine systems, contributing to better patient care and outcomes.

## 11.4 ANTENNA APPLICATIONS IN INDUSTRIAL AND MEDICAL FIELDS

### 11.4.1 Industrial applications

In industries, antennas are used for various purposes, including:

- **Process Control and Automation:** Antennas enable wireless communication between sensors, actuators, and control systems, facilitating real-time monitoring and control of industrial processes.
- **Asset Tracking and Management:** Antennas help in tracking the movement and location of assets, equipment, and vehicles within industrial facilities, improving inventory management and resource allocation.
- **Wireless Sensor Networks (WSN):** Antennas are crucial components in WSNs, which are used for monitoring environmental conditions, equipment performance, and worker safety in industrial settings.
- **Machine-to-Machine (M2M) Communication:** Antennas facilitate direct communication between machines, enabling them to share data, coordinate tasks, and optimize performance.

### 11.4.2 Medical applications

In the medical field, antennas are utilized in various applications, such as:

- **Medical Imaging:** Antennas are employed in medical imaging systems like MRI, CT, and X-ray, where they help in transmitting and receiving radio waves to generate detailed images of the human body.
- **Patient Monitoring:** Antennas are used in devices like ECG, EEG, and pulse oximeters to monitor vital signs and transmit data wirelessly to healthcare professionals for analysis and decision-making.
- **Telemedicine and Remote Healthcare:** Antennas play a vital role in enabling wireless communication between patients, doctors, and healthcare providers, facilitating remote consultations, diagnosis, and treatment.
- **Medical Robotics and Assistive Devices:** Antennas are integrated into medical robots and assistive devices to enable wireless communication and control, enhancing their functionality and effectiveness in healthcare settings.

## 11.5 OBJECTIVES AND SCOPE OF THE CHAPTER

The primary objectives of this chapter are to provide an understanding of antenna types, designs, and their applications in the industrial and medical fields. We will discuss various antenna classifications, design considerations, and advanced technologies that enhance their performance in these specific domains. Additionally, we will explore antenna integration strategies, challenges, and future trends in antenna development. By the end of this chapter, readers should have a comprehensive understanding of the importance of antennas in industrial and medical applications and their role in enhancing wireless communication.

## 11.6 ANTENNA DESIGN CONSIDERATIONS FOR INDUSTRIAL APPLICATIONS

- Industrial Wireless Communication Standards
- Antenna Gain and Directivity
- Antenna Efficiency and Radiation Patterns
- Antenna Integration with Industrial Equipment

Key considerations for designing antennas for industrial applications include:

- Compliance with relevant wireless communication standards
- Optimizing gain and directivity for efficient signal transmission and reception
- Maximizing efficiency to minimize power losses
- Designing radiation patterns to suit specific communication requirements and coverage areas

### 11.6.1 Industrial wireless communication standards

The key industrial wireless communication standards include IEEE 802.11 series (Wi-Fi) and Bluetooth Low Energy (LE). These standards impact antenna design for industrial applications by considering factors such as frequency, impedance, physical obstacles, noise level, line of sight, and topographic factors. Antenna design must account for these aspects to ensure effective wireless communication in industrial settings.

### 11.6.2 Antenna gain and directivity

Antenna gain and directivity are crucial factors in determining the performance of an antenna in an industrial setting. Gain refers to the ability of an antenna to amplify the incoming signal, while directivity measures the antenna's ability to focus the transmitted or received signal in a specific direction. High gain and directivity are essential for achieving better signal strength, coverage, and communication reliability in industrial environments. Designers should optimize these parameters to meet the specific requirements of the application, considering factors such as the size of the coverage area, the number of communication nodes, and the desired signal-to-noise ratio.

### 11.6.3 Antenna efficiency and radiation patterns

Antenna efficiency, on the other hand, is a measure of the antenna's ability to convert input power into radiated power. Efficiency is crucial in industrial applications as it directly impacts power consumption and overall system performance. Higher efficiency means less power is lost in the antenna, resulting in lower energy consumption and potentially longer battery life for wireless devices. Designers should aim to maximize antenna efficiency by minimizing energy losses due to factors such as ohmic losses in the conductors, dielectric losses in the materials, and radiation losses.

Radiation patterns, or the spatial distribution of the radiated power, also play a significant role in antenna performance. The radiation pattern should be designed to suit the specific communication requirements and coverage area. For example, in a point-to-point communication link, a highly directional antenna with a narrow beamwidth is preferred to minimize signal loss and improve the link budget. In contrast, a wide beamwidth antenna is more suitable for point-to-multipoint communication, where the antenna needs to cover a broader area.

In summary, optimizing antenna gain, directivity, efficiency, and radiation patterns is crucial for designing effective antennas for industrial applications. These factors directly impact communication reliability, power consumption, and overall system performance, making them essential considerations for antenna designers.

### 11.6.4 Antenna integration with industrial equipment

The integration of antennas with industrial equipment can affect their efficiency and radiation patterns by introducing impedance mismatches, which can lead to reflection loss and decreased radiation efficiency. Additionally, the physical placement and orientation of the antenna within the industrial equipment can impact the radiation pattern, potentially causing deviations from the desired radiation characteristics. It is important to carefully design and integrate antennas with industrial equipment to optimize their performance and ensure reliable wireless communication.

In summary, the integration of antennas with industrial equipment is crucial for reliable wireless communication in industrial settings. To achieve this, designers must consider factors such as antenna gain, directivity, efficiency, and radiation patterns. Additionally, following industrial wireless communication standards, can help ensure proper antenna design and implementation, leading to improved system performance and overall efficiency.

The industrial wireless communication standards typically include the use of IEEE 802.15.4-2006 as the physical layer, operating in the 2.4 GHz Industrial Scientific and Medical radio frequency band using 15 different channels. This diagram also shows how WirelessHART, based on the Open Systems Interconnection model, is utilized for reliable and regulatory-approved connectivity in industrial applications.

## 11.7 LITERATURE REVIEW

The current literature on Industrial Wireless Communication Standards highlights the importance of reliable and real-time communication for industrial applications. One key finding is the evaluation of Digital Enhanced Cordless Telecommunications – Ultra Low Energy (DECT-ULE) for robust communication in dense Wireless Sensor Networks (WSNs). Additionally, the IEEE 802 wireless standards, such as IEEE 802.11, play a crucial role in ensuring uniformity and interoperability in wireless network access points. While specific diagrams may not be mentioned in the snippets provided, the information emphasizes the significance of standards like IEEE 802.11 for industrial communication.

1. Zhang, H., Zhou, Y., & Liu, J. (2018). A Survey on Industrial Wireless Communication Standards. *IEEE Communications Surveys & Tutorials*, 20(1), 1215–1238. doi: 10.1109/COMST.2017.2765746.
   This paper provides an overview of various industrial wireless communication standards, including IEEE 802.15.4, WirelessHART, ISA100.11a, and Zigbee, discussing their characteristics, applications, and challenges.
2. "Antennas for All Applications – Solution Manual" by John D. Kraus: This manual is a comprehensive resource for understanding antenna design and optimization, with a focus on applications across various industries, including medical and industrial sectors. It offers detailed explanations and solutions to problems related to antenna design, making it a valuable reference for professionals and students in the field.
3. "Antennas: For All Applications, Third Edition – Instructors' Manual": This manual is an updated version of the popular textbook "Antennas: For All Applications." It serves as a guide for educators to teach antenna design and optimization concepts, including those relevant to industrial and medical applications. The manual provides additional resources and teaching materials for instructors to help students grasp the subject matter effectively.
4. "Design and Optimization of Antenna Parameters for Medical Applications": This research paper delves into the specifics of optimizing antenna parameters for medical applications, such as wearable devices and implantable systems. It discusses the challenges and requirements of antenna design in this context and presents innovative solutions to improve wireless communication in medical settings.

5. "Wearable Antennas for Biomedical Systems": This journal article focuses on wearable antennas designed for biomedical systems, including their design, optimization, and performance evaluation. It explores the potential benefits of such antennas in medical applications, such as remote patient monitoring and wireless communication between medical devices.
6. "Smart Antennas: Fundamentals and Practical Information": This book provides a comprehensive understanding of smart antenna systems, which are designed to improve wireless communication by adaptively controlling the radiation pattern of antennas. The book covers various aspects of smart antenna design, implementation, and applications, including those relevant to the industrial and medical sectors.

## 11.8 INTRODUCTION TO ANTENNAS IN CONNECTIVITY

### 11.8.1 Antenna basics and importance

Antennas are the backbone of wireless communication, translating electrical signals into radio waves and vice versa. Their performance and design directly influence connectivity range, data rate, and reliability. In critical sectors like industry and medicine, optimized antennas guarantee seamless integration and communication for devices and systems.

### 11.8.2 Evolving connectivity needs

With the growth of the IoT and smart systems, there is an ever-increasing demand for robust wireless communication. Antennas in this context are evolving to meet the needs of higher frequency bands, lower power consumption, and larger data capacities, all while maintaining signal integrity in complex environments.

## 11.9 INDUSTRIAL APPLICATIONS OF OPTIMIZED ANTENNAS

### 11.9.1 Streamlining manufacturing processes

#### *11.9.1.1 Real-time data transfer*

Optimized antennas facilitate real-time data transfer in manufacturing environments, enabling predictive maintenance and process automation. By ensuring strong and reliable signals, downtime is reduced, and efficiency is increased.

#### 11.9.1.2 Equipment connectivity

Seamless equipment connectivity is achieved with high-performance antennas tailored to withstand harsh industrial conditions. This connectivity ensures that machines communicate effectively, optimizing workflow and safety.

### 11.9.2 Enhancing automation and control

#### 11.9.2.1 Automated guided vehicles

High-precision antennas play a critical role in the navigation and operation of AGs, allowing for smooth movement and coordination in dynamic industrial environments.

#### 11.9.2.2 Remote monitoring systems

Robust antenna designs contribute to the effectiveness of remote monitoring systems, providing consistent data communication and enabling proactive management of industrial operations.

## 11.10 ANTENNA SOLUTIONS IN MEDICAL SCENARIOS

### 11.10.1 Patient monitoring and data transmission

#### 11.10.1.1 Wearable medical devices

Antennas integrated into wearable medical devices allow for continuous patient monitoring, transmitting vital data in real time to healthcare professionals, thereby facilitating timely intervention.

#### 11.10.1.2 Medical imaging equipment

Antennas in medical imaging equipment need to maintain high signal integrity to ensure accuracy and reliability in diagnostics, improving patient outcomes.

### 11.10.2 Telemedicine and remote care applications

#### 11.10.2.1 Secure data exchange

Medical antennas are designed to provide secure and reliable data exchange for telemedicine applications, safeguarding patient privacy and enhancing care delivery.

#### *11.10.2.2 Emergency response connectivity*

In emergency medical services, reliable antenna connectivity in communication devices can mean the difference between life and death, highlighting the necessity for optimized antenna function in these scenarios.

## 11.11 DESIGNING FOR RELIABILITY AND PERFORMANCE

### 11.11.1 Custom antenna design parameters

Specialized design considerations for industrial and medical antennas include frequency range, bandwidth, power handling, and environmental resilience to ensure their functionality remains uncompromised in challenging settings.

### 11.11.2 Integrating emerging technologies

#### *11.11.2.1 5G and beyond*

The integration of 5G and upcoming technologies into antenna design anticipates the need for faster data rates and lower latencies, proving essential for future-ready applications in industry and medicine.

#### *11.11.2.2 Smart antenna systems*

Smart antennas, employing multiple elements that work together, adapt their patterns of propagation to the environment, significantly enhancing communication efficiency and reliability.

## 11.12 CASE STUDIES: SUCCESS STORIES

### 11.12.1 Transforming industrial operations

Real-world examples demonstrate how optimized antennas have revolutionized industries by improving connectivity, reducing waste, and ensuring safety in complex and hazardous environments.

### 11.12.2 Innovations in healthcare and telemetry

Illustrations of how advanced antenna solutions have enabled groundbreaking progress in patient care, from remote monitoring to cutting-edge diagnostic tools, highlighting the direct impact on patient health and medical workflows.

### 11.12.3 Future trends and challenges

The penultimate section outlines emerging trends and anticipated challenges in the dynamic landscape of wireless communication for industrial and medical applications. It explores the integration of Industry 4.0 and the IoT and highlights areas for future research and development.

## 11.13 CONCLUSION

In the concluding remarks, the chapter synthesizes key findings and emphasizes the pivotal role of optimized antennas in facilitating enhanced wireless communication. It encapsulates the overarching themes discussed throughout the chapter and serves as a call to action for continued innovation in the design and optimization of antennas for industrial and medical applications.

## 11.14 CLOSING NOTE

In a world where connectivity is the lifeline of innovation and progress, optimizing antenna technology in industrial and medical applications is not just an upgrade—it's a revolution. Let's embrace this wireless future, ensuring reliability and superior performance that save time, reduce costs, and ultimately save lives is shown in Figure 11.1.

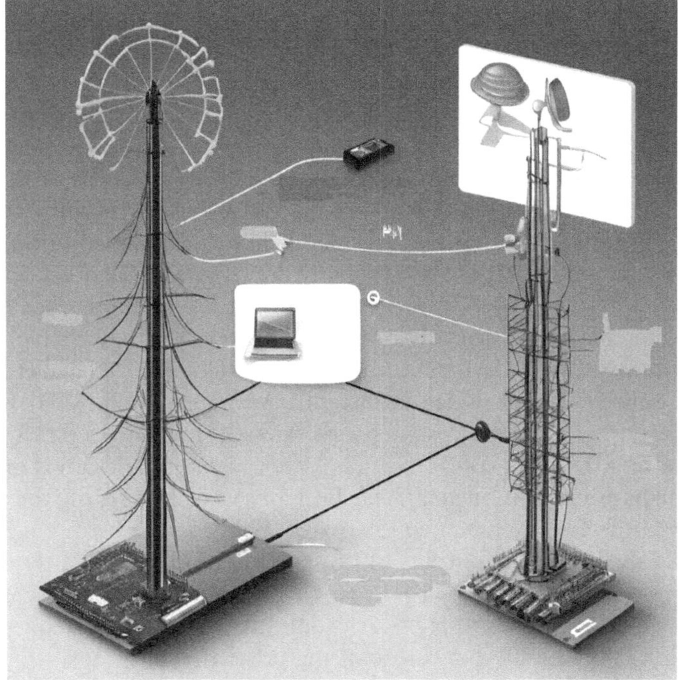

*Figure 11.1* Wireless communication.

Chapter 12

# Design of 3 GHz rectangular microstrip patch antenna and modeling using surface-fitting technique

*Shakil Ahmed and Sudip Mandal*

## 12.1 INTRODUCTION

Microstrip antennas have extensively been used in commercial and military communication systems due to their simplicity, conformability, low manufacturing cost, lightweight, low profile, reproducibility, reliability, and ease of fabrication and integration with solid-state devices [1].

A Microstrip patch antenna consists of a radiating patch on one side of a dielectric substrate which has a ground plane on the other side. The patch is generally made of conducting material such as copper or gold and can take any possible shape. The radiating patch and the feed lines are usually photo-etched on the dielectric substrate [2]. The patch may be in a variety of shapes, but rectangular and circular are the most common because of ease of analysis and fabrication, attractive radiation characteristics, and especially low cross-polarization. The modes are supported primarily by a rectangular microstrip patch antenna whose substrate height is small [3].

The use of sub 3 GHz frequencies combined with recent technological advances has resulted in the reach of digital terrestrial radio systems being greatly extended [4]. Key advantages of sub 3 GHz transmission include largely interference-free operation in licensed bands; an inherent long-distance capability; and immunity to many atmospheric conditions. Modest infrastructure requirements also result in lower capital and operational costs, leading to a low total cost of ownership. 3 GHz RF frequency has typically some interesting applications like IoT applications, body wearable antenna for health monitoring, ISM communication, WLAN etc. [5–10] dictating the efficiency and reliability of data transfer between interconnected devices.

In this chapter, we have designed a rectangular microstrip patch antenna operating at a frequency of 3 GHz. The patch antenna is designed on a FR4. Next, the parametric influences have been observed by varying the antenna design parameters like patch length, patch width, and substrate height and observing the respective outputs like $S_{11}$. CST studio [11] has been used for this purpose. In the next phase of this work, a surface fitting model has been proposed that helps to predict the output $S_{11}$, resonance frequency, and bandwidth. The results show a promising predicting accuracy.


DOI: 10.1201/9781003560487-12

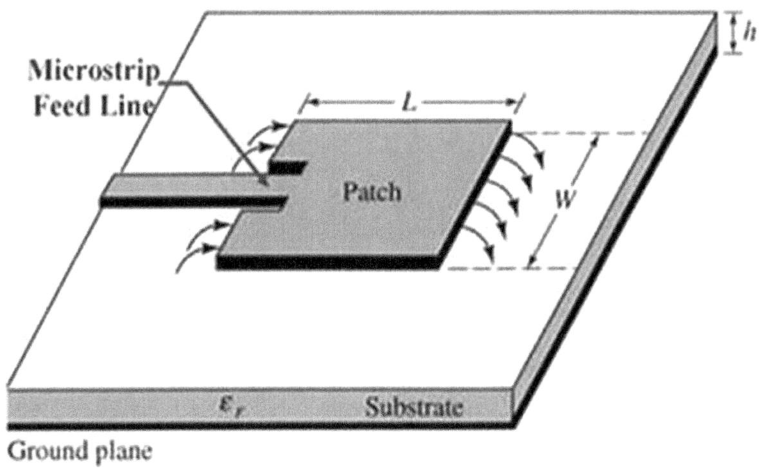


*Figure 12.1* CST design of proposed antenna [1].

## 12.2 METHODOLOGY

The geometry of the rectangular microstrip antenna is shown in Figure 12.1. The length and width of the rectangular microstrip antenna are calculated by using the below formulas (Eqs. 12.1–12.7).

The length and width of a rectangular antenna may be evaluated through the usage of the subsequent equations [12–14]:

$$W = \frac{C}{2f_r}\sqrt{\frac{2}{\varepsilon_r + 1}} \tag{12.1}$$

where

$C$ is speed of light,
$f_r$ is the resonant frequency,
$\varepsilon_r$ is the relative permittivity of the substrate

$$L = L_{\text{eff}} - 2\Delta L \tag{12.2}$$

where

$L_{\text{eff}}$ is the effective length of the patch can be expressed as

$$L_{\text{eff}} = \frac{c}{2f_r\sqrt{\varepsilon_{\text{eff}}}} \tag{12.3}$$

where

$\varepsilon_{\text{eff}}$ is the effective relative permittivity given as follows:

$$\varepsilon_{\text{eff}} = \frac{\varepsilon_r + 1}{2} + \frac{\varepsilon_r - 1}{2}\left(1 + \frac{12h}{W}\right)^{-\frac{1}{2}} \tag{12.4}$$

and $\Delta L$ is the length extension given as follows

$$L = 0.412h \frac{(\varepsilon_{\text{eff}} + 0.3)\left(\frac{W}{h} + 0.264\right)}{(\varepsilon_{\text{eff}} - 0.258)\left(\frac{W}{h} + 0.8\right)} \tag{12.5}$$

The rectangular patch antenna's dimensions (length×width×height) are roughly indicated by the matching length $L_g$, width $W_g$, and height of the substrate $h$. These values are determined using the provided equations.

$$L_g = L + 6h \tag{12.6}$$

$$W_g = W + 6h \tag{12.7}$$

The proposed methodology consists of three stages which are described one by one below.

### 12.2.1 Schematic and design of proposed microchip patch antenna

We have designed a rectangular microstrip patch antenna using the CST Studio antenna design tool, which will work on a center frequency of 3 GHz. We used annealed copper in the patch and ground plate. For better gain, the substrate is chosen as FR-4(lossy) with an epsilon value of 4.3 and loss tangent can vary from 0.015 to 0.025. We have used the inset feed method for impedance matching where the characteristic impedance is assumed to be 50 Ω. Initially, the rough dimensions of the rectangular patch antenna have been calculated using the conventional equations with 1 mm thick substrate and 0.035 mm thick copper patch.

Then manual tuning of different dimensions i.e. length, width of rectangular patch, and substrate depth ($D_s$) is done to achieve the best performance of the antenna. The outline structure of the proposed antenna (using CST studio) is given in Figure 12.2. The values of the parameters for the best result after tuning are listed below in Table 12.1 in tabular form.

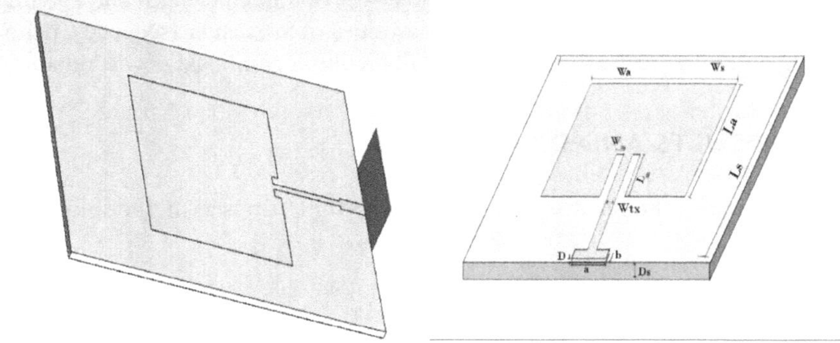


*Figure 12.2* Schematic of proposed 3 GHz rectangular patch antenna.

*Table 12.1* Parameters for the best result of the proposed antenna

| *Sl. No.* | *Parameter* | *Value (mm)* |
|---|---|---|
| 1 | Antenna length ($L_a$) | 22.55 |
| 2 | Antenna width ($W_a$) | 32 |
| 3 | Substrate depth ($D_s$) | 1.7 |
| 4 | Substrate width ($W_s$) | 50 |
| 5 | Substrate length ($L_s$) | 50 |
| 6 | Patch depth ($D$) | 0.035 |
| 7 | Transmission line width ($W_{tx}$) | 3.333 |
| 8 | Inset length ($L_{in}$) | 1 |
| 9 | Inset width ($W_{in}$) | 1 |
| 10 | *A* | 4 |
| 11 | *B* | 2.667 |

### 12.2.2 Observation of parametric influence

Now, to observe the influence of various rectangular patch antenna parameters (i.e. length, width, height etc.) on the outputs (i.e. $S_{11}$), we have changed the parameter values with some specified step variations. Then, we ran the simulation for each variation, plotted them together in one graph for easier understanding, and observed the influences on the outputs.

### 12.2.3 Patch antenna modeling using surface fitting

Next, we develop a mathematical model with which we can predict the values of $S_{11}$, center frequency, and bandwidth using the surface fitting technique. If we consider substrate height 1.7 and put the values of antenna length and antenna width in this model, then we can get the values of $S_{11}$, center frequency, bandwidth, and other parameters automatically without running a new simulation.

## 12.3 RESULTS AND DISCUSSIONS

In this section, the results corresponding to the proposed methodology are described.

### 12.3.1 Outputs of optimized 3 GHz rectangular patch antenna

By adjusting the different dimensions of the proposed rectangular patch antenna, initially, we have tried to reach the best case i.e. close to the 3 GHz operating frequency with minimum $S_{11}$ value. The respective dimensions of the antenna parameters were already given in Table 12.1.

#### *12.3.1.1 $S_{11}$ curve for best case scenario*

Figure 12.3 shows the variation $S_{11}$ value with respect to frequency for the proposed antenna. It has been observed that the proposed antenna resonates at 2.998 GHz frequency and the corresponding $S_{11}$ value is –36.71 dB. Our targeted center frequency of the antenna was 3 GHz and achieved simulated frequency was 2.998 GHz which is very close to the desired value. Also the corresponding $S_{11}$ value or return loss is –36.71 dB which indicates a very satisfactory finding that indicates a very good matching. Table 12.2 summarizes the performance of the proposed antenna in terms of $S_{11}$, operating frequency, and bandwidth.

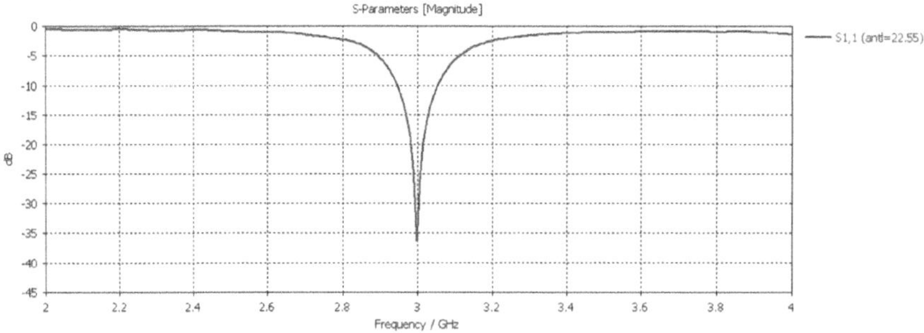


*Figure 12.3* Variation of $S_{11}$ vs. frequency.

*Table 12.2* Return loss and frequencies for proposed patch antenna

| *Operating frequency* | *Return loss* | *Lower frequency* | *Upper frequency* | *Band width* |
|---|---|---|---|---|
| 2.998 GHz | −36.71 dB | 2.9461 GHz | 3.0546 GHz | 0.1085 GHz |

#### *12.3.1.2 Bandwidth*

From Figure 12.3, upper frequency and lower frequency (–10 dB less) of the proposed rectangular patch antenna are found at 3.0546 and 2.9461 GHz, respectively. Hence, the bandwidth of the antenna is found as (3.0546–2.9461) GHz = 0.1085 GHz i.e. 108.5 MHz. It is very satisfactory for the IoT communication.

#### *12.3.1.3 Gain*

Figure 12.4 shows the variation gain of our proposed antenna with respect to its frequency. It has been observed that up to frequency 3 GHz (which is our desired center frequency) the gain rises linearly and at 3 GHz frequency the gain value is 2.76 dB (satisfactory level). After 3 GHz frequency, the gain remains almost constant.

#### *12.3.1.4 VSWR*

Figure 12.5 shows the Voltage Standing Wave Ratio (VSWR) curve. It has been observed that the proposed antenna least reflects at 2.998 GHz frequency and the corresponding VSWR value is nearly 1.026 which indicates a very satisfactory matching of the proposed antenna.

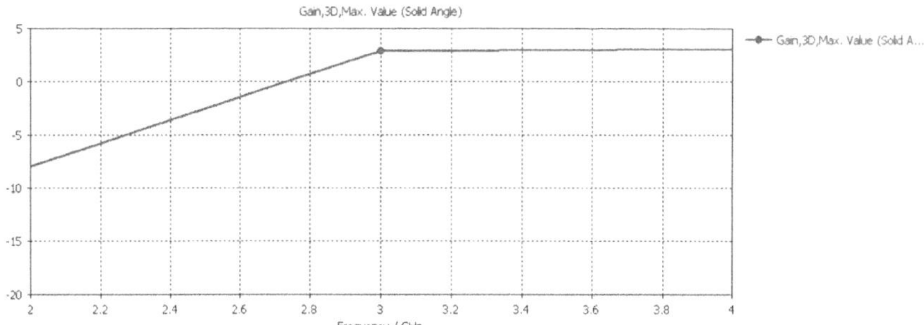


*Figure 12.4* Variation of gain vs. frequency.

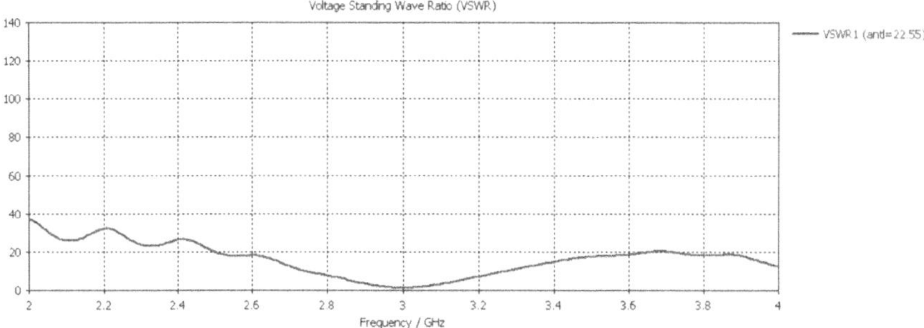


*Figure 12.5* Variation of VSWR vs. frequency.

#### *12.3.1.5 Impedance*

Figure 12.6 shows Z-parameters or Impedance Parameters with respect to frequency. We know the antenna radiates maximum power when the impedance is 50 Ohm. It has been clearly observed that the impedance at 2.998 GHz is around 48.3 Ohm which is close to the characteristics impedance due to the utilization of the inset feeding technique.

#### *12.3.1.6 Radiation pattern*

Figure 12.7 shows far-field radiation pattern of the rectangular patch antenna. It has been observed that the HPBW of the proposed microstrip patch antenna is 94.2° which is good for the IoT communication.

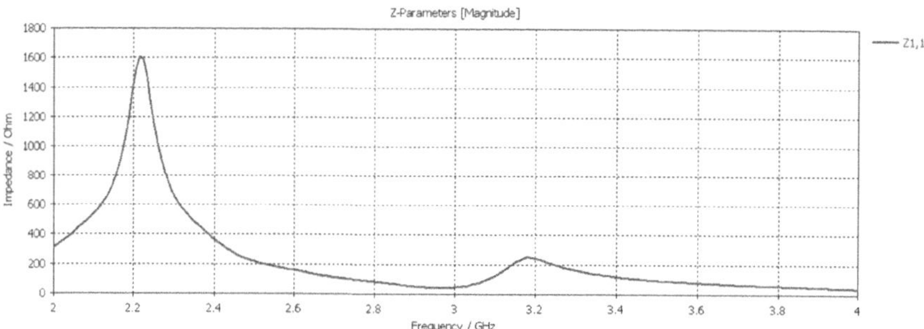


*Figure 12.6* Variation of impedance vs. frequency.

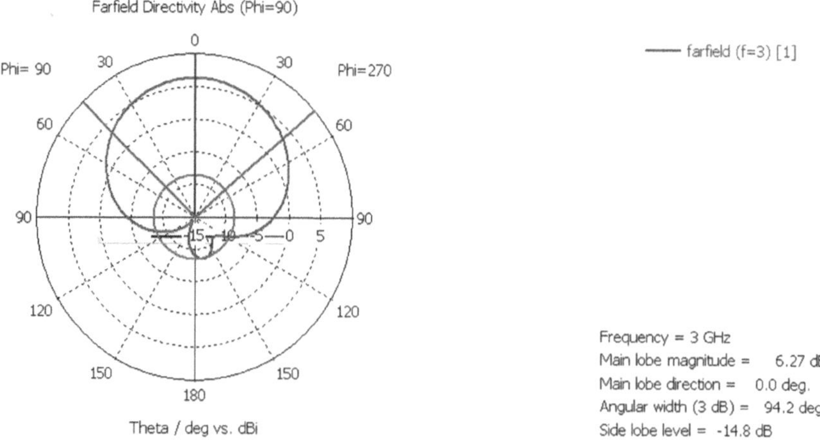


*Figure 12.7* Radiation pattern of proposed antenna.

*Table 12.3* Comparative study with similar works

| *Work* | $S_{11}$ | *Bandwidth (MHz)* | *Gain (dB)* | *HPBW* |
|---|---|---|---|---|
| Verma et al. [6] | −18 dB | 3.0400–2.9800 = 60 | 3.98 | – |
| Singhal et al. [8] | −26 dB | 3.0250–2.9550 = 70 | 6.50 | – |
| Proposed work | **−36.71 dB** | 3.0546–2.9461 = **108.5** | 2.76 | 94.2° |

#### *12.3.1.7 Comparative study*

Here, the outputs (i.e. $S_{11}$, bandwidth, Gain, and HPBW) of the proposed 3 GHz rectangular patch antenna are compared with similar works of MPA designing at 3 GHz.

From Table 12.3, we can observe that the proposed rectangular patch antenna is superior to other two similar works Verma et al. [6] and Singhal and Jaimini [8] with respect to $S_{11}$ and bandwidth (i.e. –36.71 dB and 108.5 MHz, respectively). The best values are shown with bold front. However, the gain of the proposed antenna is slightly less compared to the other techniques. HPBW for other works was not reported.

### 12.3.2 Results for parametric influences

Now, we have observed the variation of $S_{11}$ parameter with respect to the frequency for different values of length and width of the patch. While varying one parameter, other dimensions are kept unchanged i.e. remained in the same values as mentioned in Table 12.1. The following three cases are considered.

#### *12.3.2.1 Variation of length*

In this scenario, we altered the antenna's length. We tested seven different values, decreasing the length three times with a 1 mm step and increasing it three times with a 1 mm step from the optimal value (22.55 mm). For each of these seven length values, we simulated the antenna using CST Studio, observed changes in the $S_{11}$ parameter, and illustrated the results in the graphs displayed in Figure 12.8.

From Figure 12.8, the optimal $S_{11}$ values for lengths of 19.55 mm, 20.55 mm, 21.55 mm, 22.55 mm, 23.55 mm, 24.55 mm, and 25.55 mm are –26 dB, –28 dB, –29 dB, –37 dB, –36 dB, –31 dB, and –25 dB, respectively. It is evident that the most favorable $S_{11}$ value occurs at a length of 22.55 mm, measuring almost –37 dB at a frequency of 2.998 GHz, aligning with our expectations. Analyzing the graph, we observe that as the antenna length increases, the frequency at which $S_{11}$ is at its minimum decreases, indicating a reduction in the center frequency. However, the overall pattern of

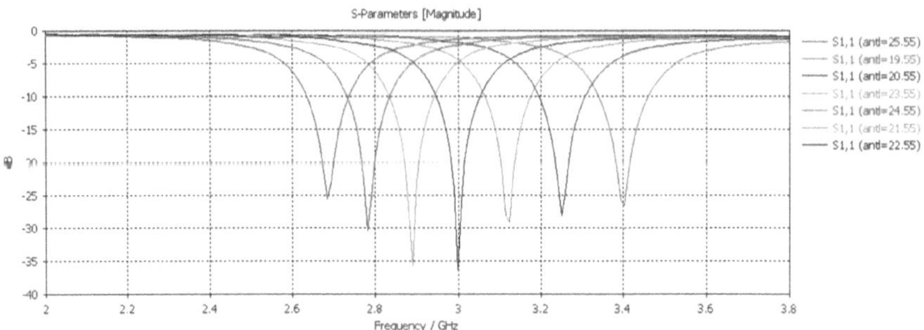


*Figure 12.8* Influence of length on $S_{11}$ parameter.

the graph remains largely unchanged. Therefore, we can conclude that the graph shifts to the left with an increase in antenna length.

#### *12.3.2.2 Variation of width*

In the second case, we varied the width ($w$) of the antenna. We tested seven different values of the substrate width, decreasing the value three times with a step difference of 1 mm and increasing it three times with a step difference of 1 mm from the optimal value (32 mm). For each of these seven width values, we simulated the antenna in CST Studio, observed changes in the $S_{11}$ parameter, and depicted the results in the graphs shown in Figure 12.9.

From Figure 12.9, the optimal $S_{11}$ values for substrate widths of 29 mm, 30 mm, 31 mm, 32 mm, 33 mm, 34 mm, and 35 mm are –24 dB, –31 dB, –33 dB, –37 dB, –28 dB, –24 dB, and –23 dB, respectively. It is evident that the most favorable $S_{11}$ value occurs at a width of 32 mm, measuring –36.71 dB at a frequency of 2.998 GHz, as expected. Analyzing the graph, we observe that with an increase in substrate width up to 32 mm (the best case), the $S_{11}$ value increases. However, beyond 32 mm, the $S_{11}$ value starts to decrease again.

#### *12.3.2.3 Variation of substrate height*

In the second scenario, we manipulated the substrate height of the antenna. We examined five different values of the substrate height, reducing the value two times with a step difference of 0.2 mm and increasing it two times with a step difference of 0.2 mm from the optimal value (1.7 mm). For each of these seven substrate depth values, we conducted antenna simulations in CST Studio, monitored variations in the $S_{11}$ parameter, and presented the outcomes in the graphs displayed in Figure 12.10.

In Figure 12.10, the optimal $S_{11}$ values for depth of 1.3 mm, 1.5 mm, 1.7 mm, 1.9 mm and 2.1 mm are –22 dB, –27 dB, –42 dB, –29 dB and –23 dB, respectively. It is evident that the most favorable $S_{11}$ value occurs at a depth

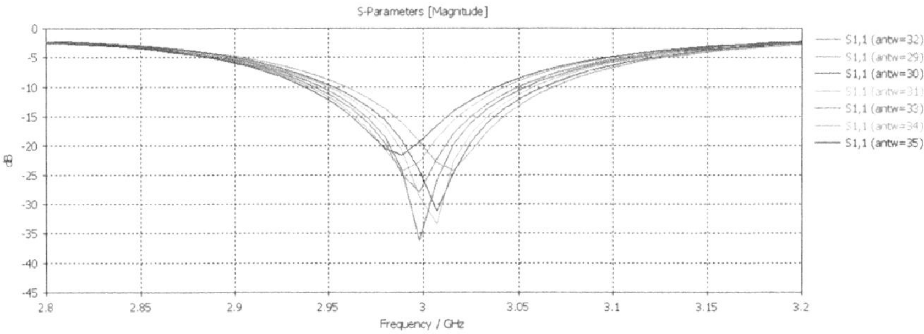


*Figure 12.9* Influence of width on $S_{11}$ parameter.

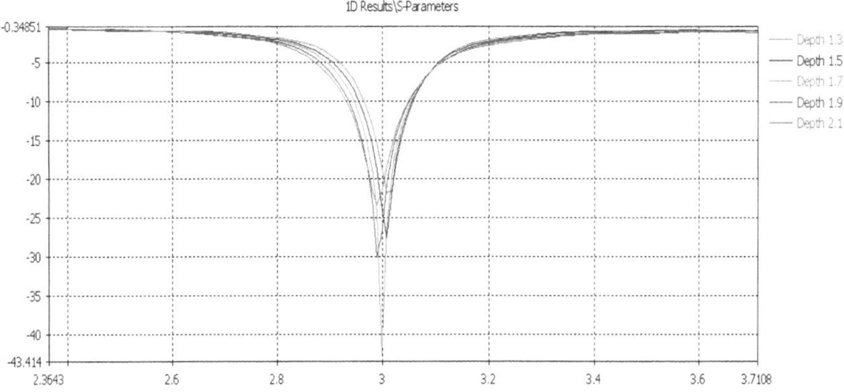


*Figure 12.10* Influence of substrate height on $S_{11}$ parameter.

of 1.7 mm. Analyzing the graph, we observe that with an increase in height up to 1.7 mm (the optimal case), the $S_{11}$ value increases. However, beyond 1.7 mm, the $S_{11}$ value starts to decrease again.

### 12.3.3 Surface fitting model of proposed 3 GHz circular patch antenna

We are currently working on establishing a mathematical model that enables the prediction of $S_{11}$ values, center frequency, bandwidth, and other parameters. Once developed, this model will allow us to input specific values for the antenna length and antenna width, providing automatic predictions for $S_{11}$, center frequency, bandwidth, and other relevant parameters without the need for simulation runs. This approach aims to streamline the prediction process and enhance efficiency in obtaining key antenna performance metrics.

For this research, we systematically varied two fundamental parameters of the antenna i.e. length and width across seven discrete settings, precisely collecting a dataset comprising 49 data points. This hands-on method made it easier to identify the crucial measures that determine how well something is performing, including the $S_{11}$ value, center frequency, and bandwidth. The deliberate manipulation of these antenna dimensions allowed for a comprehensive analysis of their impact on the antenna's operational characteristics across various frequencies. The findings from this investigation provide a valuable framework for optimizing antenna design, enabling tailored enhancements to achieve superior performance within specific frequency ranges. This research offers valuable insights and guidance for future advancements in antenna engineering and design optimization.

Now, we utilized the MATLAB® surface fitting tool to fit the values of various parameters obtained from Excel sheets into a curve. For precise fitting, we employed a polynomial of degree 5, denoted as poly55, in our curve fitting. The generalized equation for this polynomial is as follows:

$$\begin{aligned}\text{surffit}(x,y) = {} & p00 + p10 * x + p01 * y + p20 * x^2 + p11 * x * y \\ & + p02 * y^2 + p30 * x^3 + p21 * x^2 * y + p12 * x * y^2 + p03 * y^3 \\ & + p40 * x^4 + p31 * x^3 * y + p22 * x^2 * y^2 + p13 * x * y^3 \\ & + p04 * y^4 + p50 * x^5 + p41 * x^4 * y + p32 * x^3 * y^2 \\ & + p23 * x^2 * y^3 + p14 * x * y^4 + p05 * y^5\end{aligned} \tag{12.8}$$

where $x$ is normalized by mean 32 and std 2.021 and where y is normalized by mean 22.55 and std 2.021. Here, $x$ and $y$ are input variable, surffit is the output parameters, ($p00$, $p10$,..., $p55$) are the respective coefficients of the polynomial.

#### 12.3.3.1 Modeling of $S_{11}$

Using surface fitting, we have obtained the following polynomial function to fit the best $S_{11}$ parameter with respect to length ($L_a$) and width ($W_a$). The coefficients for this surface fitting model is given as: $p00=32.83$, $p10=-6.337$, $p01=0.9111$, $p20=-4.283$, $p11=8.643$, $p02=-2.219$, $p30=2.631$, $p21=-2.989$, $p12=3.078$, $p03=0.5694$, $p40=-0.1617$, $p31=-2.446$, $p22=1.443$, $p13=-0.3892$, $p04=0.04057$, $p50=-0.1606$, $p41=0.7063$, $p32=-0.9327$, $p23=0.7574$, $p14=-0.2012$, $p05=-0.4242$

The obtained surface fitting plot for best $S_{11}$ is shown in Figure 12.11.

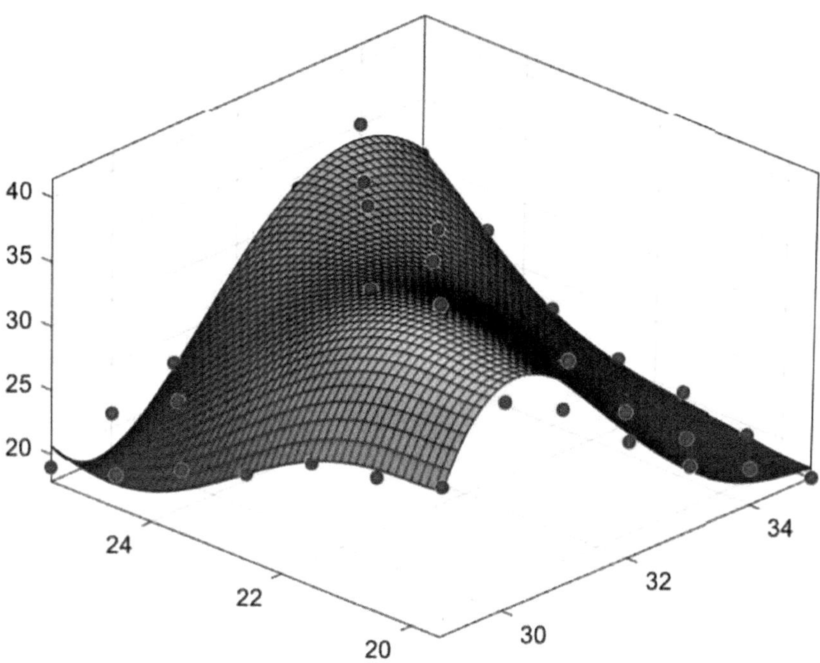


*Figure 12.11* Surface fitting of $S_{11}$ with respect to $L_a$ and $W_a$.

#### *12.3.3.2 Modeling of center frequency*

Using surface fitting, we have obtained the following polynomial function to fit the center frequency parameter with respect to length ($L_a$) and width ($W_a$). The coefficients for this surface fitting model is given as $p00$=3.001, $p10$=–0.0082, $p01$=–0.2338, $p20$=–0.002777, $p11$=0.000901, $p02$= 0.01743, $p30$=0.000835, $p21$=3.684e-05, $p12$=–0.002206, $p03$= –0.002726, $p40$=0.00138, $p31$=–0.0001489, $p22$=0.0002977, $p13$=–0.0005955, $p04$=0.0008932, $p50$=–0.000361, $p41$=0.0004512, $p32$=0.0005014, $p23$=–0.0006017, $p14$=0.0005743, $p05$=–0.0001805

The obtained surface fitting plot for center frequency is shown in Figure 12.12.

#### *12.3.3.3 Modeling of bandwidth*

Using surface fitting, we have obtained the following polynomial function to fit the bandwidth with respect to length ($L_a$) and width ($W_a$). The coefficients for this surface fitting model is given as: $p00$=0.1084, $p10$=–0.001846,

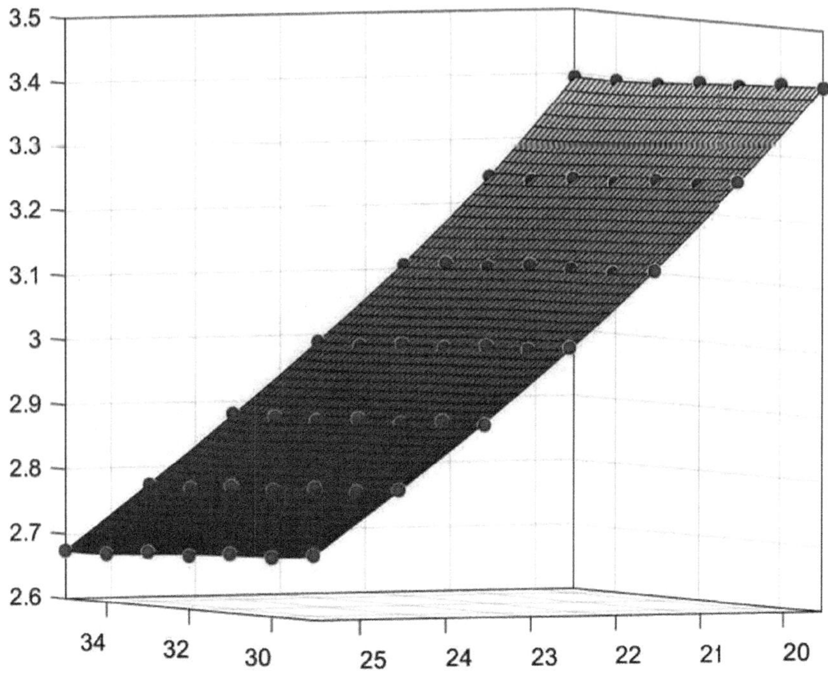


*Figure 12.12* Surface fitting of center frequency with respect to $L_a$ and $W_a$.

$p01$=–0.004065, $p20$=–0.001213, $p11$=0.001708, $p02$=0.0004171, $p30$=–0.001773, $p21$=0.0003814, $p12$=–0.0007555, $p03$=–0.00105, $p40$=0.0001173, $p31$=7.278e-05, $p22$=0.0001453, $p13$=–0.0001356, $p04$=0.000305, $p50$=0.0005816, $p41$=–6.381e-06, $p32$=0.0002206, $p23$=–8.913e-05, $p14$=0.0002853, $p05$=0.0001825

The obtained surface fitting plot for center frequency is shown in Figure 12.13.

#### *12.3.3.4 Model validation*

Now, to understand how much our curve fitting efficiently works, using the above-mentioned equations $S_{11}$, the center frequency, and bandwidth of the rectangular MPA can be calculated or predicted for a set of radius and substrate height. From there, error can be calculated from the difference between predicted and simulated values for all those 49 observations. The error can be defined as using the following formula.

$$\text{Error} = (\text{Actual value} - \text{Predicted value}) \tag{12.9}$$

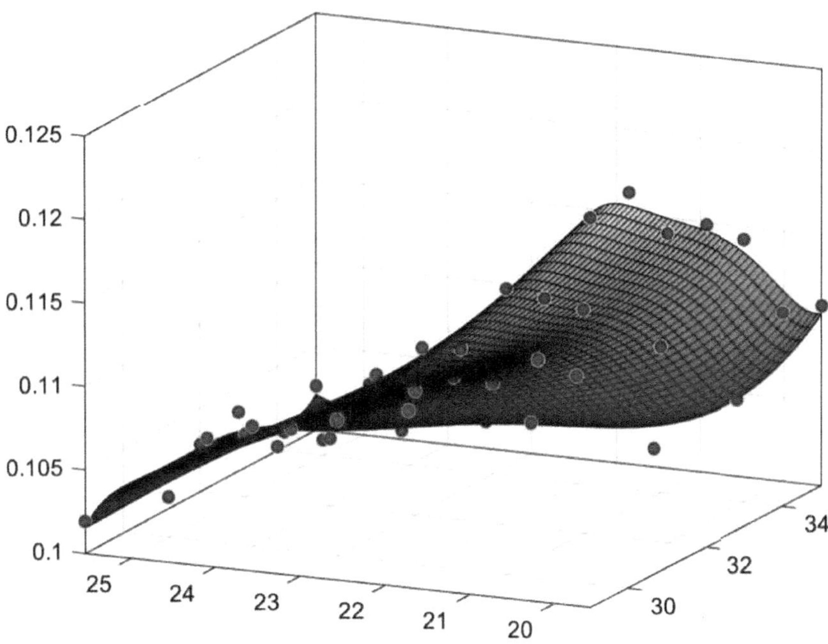


*Figure 12.13* Surface fitting of center frequency with respect to $L_a$ and $W_a$.

After collecting all simulation results, we put the values from the simulation to the mentioned equation and got the following result (Figures 12.14–12.16).

Next, Root Mean Square Error (RMSE) error for each case ($S_{11}$, center frequency, bandwidth) is determined that indicates the accuracy for prediction of the developed models. The formula for the root mean square error for $N$ observation is

$$\text{RMSE} = \sqrt{\frac{\sum_{i=1}^{N}\left(\text{Actual}_i - \text{Predicted}_i\right)^2}{N}} \tag{12.10}$$

From Table 12.4, RMSE for $S_{11}$, center frequency, and bandwidth prediction are provided. It can be observed that RMSE for $S_{11}$ is slightly high. However, the RMSE for center frequency and bandwidth is negligible which indicates that the proposed surface fitting model for the CF and BW works perfectly. This type of mathematical modeling significantly reduces the efforts and time required for the simulation of the circular patch antenna works in this 3 GHz frequency band.

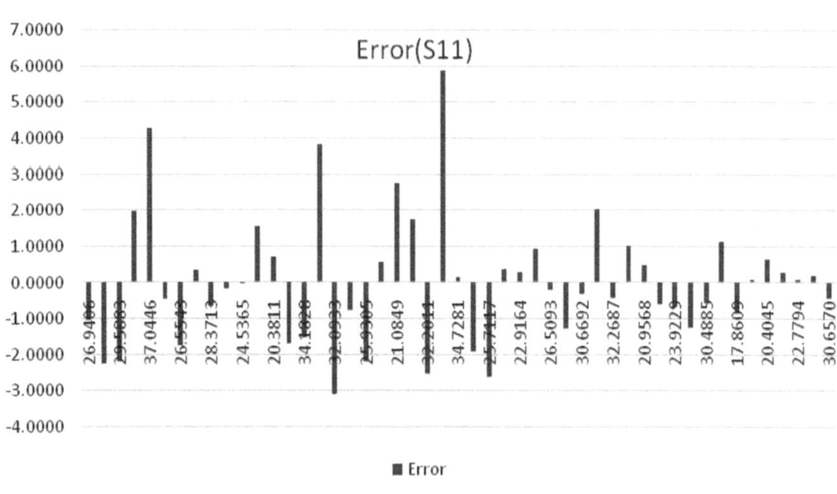


*Figure 12.14* Variation of $S_{11}$ error with different observation.

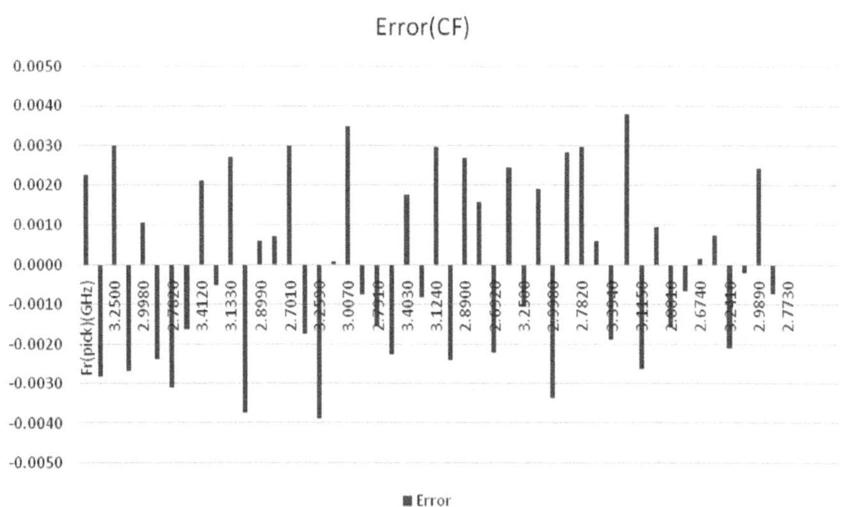


*Figure 12.15* Variation of center frequency error with different observation.

*Figure 12.16* Variation of center frequency error with different observation.

*Table 12.4* RMSE for different outputs

| *Parameter* | *RMSE* |
|---|---|
| $S_{11}$ | 1.745096744 |
| Center frequency | 0.002183666 |
| Band width | 0.000856126 |

## 12.4 CONCLUSION

In this chapter, we have designed a rectangular microstrip patch antenna operating at a frequency of 3 GHz which is suitable for IoT communication. The patch antenna is designed on a substrate FR4 and simulated using CST Studio. The output of the proposed antenna is very good with respect to $S_{11}$ and bandwidth compared to the existing works in this frequency band. Next, the parametric influences have been observed by varying the antenna design parameters (i.e. patch length and width) and observing the respective output like $S_{11}$, bandwidth, and center frequency.

In the next phase of this work, a surface fitting model has been introduced that helps to predict the output $S_{11}$, resonance frequency, and bandwidth. The results show a promising predicting accuracy and it will help to reduce the effort and time required for redesigning and simulation of this type of circular patch antenna.

## REFERENCES


1 Balanis, C.A., 2015. *Antenna theory: analysis and design.* John Wiley & Sons, New York.
2 https://www.antenna-theory.com/antennas/patches/antenna.php [accessed on 07/02/2023].
3 Singh, I. and Tripathi, V.S., 2011. Microstrip patch antenna and its applications: a survey, *International Journal of Computer Technology and Applications,* 2(5), 1595–1599.
4 *Advantages of sub 3 GHz transmission, 4RF,* 2012. www.4rf.com [accessed on 12/12/2023].
5 Kavitha, V., Ferveez, I. and Ahamed, H.A., 2017. Body-wearable flexible RF antenna for 3 GHz jamming applications, *International Journal of Current Engineering and Scientific Research,* 4(11), 16–19.
6 Verma, S., Kumar, A. and Agrawal, N., 2023. Design and optimization of a 3GHz single band patch antenna using HFSS: performance analysis and implementation for IoT, *International Journal of Intelligent Systems and Applications in Engineering,* 11(7s), 460–465.
7 Mouniga, J. and Henridass, A., 2016. Lightweight and flexible microstrip 2x2 wearable antenna array for 2.45GHz ISM application, *International Journal of Emerging Technology in Computer Science & Electronics,* 23(2), 32–36.
8 Singhal, P. and Jaimini, K., 2013. Rectangular microstrip patch antenna design at 3 GHz using probe feed, *International Journal of Emerging Technology and Advanced Engineering,* 3(11), 421–425.
9 Daliri, A., Galehdar, A., John, S., Rowe, W. and Ghorbani, K., 2010. Circular microstrip patch antenna strain sensor for wireless structural health monitoring. *Proceedings of the world congress on engineering,* 2, 1173.
10 Pandey, S. and Markam, K., 2016. Design and analysis of circular shape microstrip patch antenna for C-band applications. *International Journal of Advanced Research in Computer Science & Technology,* 4(2), 169–171.
11 CST Studio Suite, https://edu.3ds.com/en/software/cst-studio-suite-students-edition [accessed on 05/03/2022].
12 Nayak, A., Dutta, S. and Mandal, S., 2023. Design of dual band microstrip patch antenna for 5G communication operating at 28 GHz and 46 GHz, *International Journal of Wireless and Microwave Technologies,* 13(2), 43–52.
13 Dey, T. and Mandal, S., 2024. Design and analysis of 28 GHz microstrip patch antenna for 5G network, In: Tavares, J. M. R. S., Rodrigues, J. J. P. C., Misra, D., Bhattacherjee, D. (eds) *Data science and communication, ICTDsC 2023. Studies in autonomic, data-driven and industrial computing*, Siliguri, India, 773–784.
14 Sadhu, S., Acharjee, B. and Mandal, S., 2024. Design of wideband microstrip patch antenna at 38 GHz for 5G network, In: Tavares, J. M. R. S., Rodrigues, J. J. P. C., Misra, D., Bhattacherjee, D. (eds) *Data science and communication, ICTDsC 2023. Studies in autonomic, data-driven and industrial computing,* Siliguri Institute of Technology, India (during 23–24 March 2023 in Siliguri, India) 719–730.

Chapter 13

# Design of microstrip patch antenna for skin cancer detection

*N. M. Mary Sindhuja, R. Mathu Mietha, T. N. Avushatha, and V. Karthiga devi*

## 13.1 INTRODUCTION

In the field of medical applications, cancer is the primary reason for the increase in the mortality rate globally. It is a disease caused due to abnormal cell growth. Over 100 types of cancer affect humans. One among those is skin cancer which is caused mainly due to exposure to ultraviolet radiation from the sun. Skin cancer also has the danger of getting other body parts affected due to the spread of cancer. However, early identification with treatment helps human being to save their lives.

The detection techniques of skin cancer include Biopsy and certain Imaging techniques. All biopsy methods, except optical biopsy, require the removal of tissues from the affected or doubted area. Later the removed tissue undergoes microscopic examinations under chemicals and sometimes genetic tests to get nearly accurate results. This process is time-consuming, which raises the danger of infection and even death. However, certain more recent biopsies, such as reflectance co-focal microscopy, or RCM, do not require skin samples for performance. However, not all testing facilities and hospitals provide this technique, nor is it inexpensive.

Imaging techniques are done by creating pictures by the use of X-rays and other radioactive substances. Magnetic Resonance Imaging (MRI) can also be used for imaging the damaged tissue. Both techniques are not preferable because of the ionized radiations. The imaging technique requires millions of cancer cells to make a tumor big enough to be shown. Sometimes the cancer cells detected in imaging techniques turn out non-cancerous in biopsy.

In this decade for medical applications the use of antennas is increasing because of their feasibility and low power transmission. The US preventive services task force (USPSTF) has modified its stand that diagnosis of cancer using a microwave causes less health hazard with minimal discomfort and easy interpretation by considering differences in the electrical properties [1]. In recent papers RF antennas are used to detect various types of cancers. The ultra-wideband antenna with the shape of a quasi-rhomboid is used for

DOI: 10.1201/9781003560487-13

breast cancer detection. The performance of the antenna at 3.34–9 GHz exhibits return loss which is reduced to −10 dB. The antenna is connected at one end of the vector network analyzer and in other port the probe is connected to cancer cancer-affected region. The insertion loss obtained is the highest compared to other normal positions in the breast [2]. Breast cancer detection using non-ionized radiation is very cost-effective and many researchers are attracted toward this technique. Their work proves that it is an effective method of detecting cancer in various body parts [3].

The H-shaped wearable slot antenna is designed to detect cancer in the thyroid gland which is a wearable antenna. The thyroid gland model is designed and analyzed by considering with and without tumor conditions by obtaining SAR, loss, and gain values of the designed antenna [4]. The wearable antenna with center frequency of 2.5 GHz is designed using a silk substrate that takes the shape of Z. The gain value obtained is around 1.67 dB and there is a change in performance due to with and without cancer in the analysis [5]. The various antennas were proposed to detect the abnormalities in the skin as normal or tumor-based skin texture. A hexagonal patch antenna is proposed using a jean substrate operating around 2.5 GHz. The antenna in the presence and absence of a slot is designed and used to analyze the RF characteristics with and without tumor. The variation in loss, gain, and current density is obtained which reduces the risk of detecting tumor using high-intensity X-rays [6]. To reduce the stopband level and improve the bandwidth, the change in the antenna structures, size reduction, materials, and truncating the ground to $\lambda/8$. At the ISM band the antenna characteristics observed provide a decrease in radiation efficiency and coupling when the stub is added hence stub length designed should be related to resonance at the lower band. In the designed antenna the Indium Tin Oxide is used to enhance the antenna radiation properties [7]. Due to destructive rays of sunlight, the human body is susceptible to sunlight causing skin cancer. At ultra-wideband frequency, the wide band elliptical antenna is designed to work at 15 GHz with 23% of miniaturization and finds application in medical imaging and radars [8].

The monopole circular patch antenna with a rectangular slot is used to discover tumors in the skin at 2.4 GHz. The variation in RF parameters such as electric field, magnetic field, and current density are observed by simulating in CST software in the presence and absence of tumor where electric field and current density increase in the presence of tumor is observed. It is noted that more accuracy is achieved with a curved-shaped antenna which suits skin structure [9]. The PIFA antenna is used to detect skin cancer by considering and comparing values of permittivity, permeability, absorption power, and conductivity of normal and abnormal tissue. The skin tissue is placed around 10 mm distance and measured for volume loss which is different for normal and cancerous tissue [10]. The metamaterial-based wearable antenna with a resonance frequency of 11.72 GHz is developed to find skin cancer. The antenna is of star-shaped ring-structured antenna simulated in HFSS software. The size of the antenna is reduced in turn increase in gain and bandwidth is achieved due to the metamaterial used for antenna design [11]. The negative SNG (Single Negative) load-based antenna at the THz range is evolved for primal skin cancer discovery. The SNG load is used to expand the gain and frequency of the dual-band antenna which can be used to detect cancer with a circular slot in the ground plane. The metamaterial characteristics of the antenna are analyzed for various frequencies such as 0.85, 0.94, 0.98, and 1 THz by observing the current distribution, reflection coefficient and very high gain [12].

An antenna designed at 2.45 GHz is used to detect normal and cancerous tissue prepared by pathologists by inferring RF parameters such as return loss, impedance value, and *E* and *H* field pattern. The analysis is done by considering various distances between the antenna and tissue [13]. This work builds, analyzes, and attempts to fabricate a dynamically polarized antenna. The frequency range sheltered by the antenna's gain is 390–610 MHz. The polarized antenna produces an output of about 19 dBm. Voltage-controlled diodes and varactor diodes has been employed to attain steady functioning [14]. An extensive review of the most latest promotions in microwave imaging and sensing methods for early detection of breast cancer is given in this publication. It explains the fundamentals and uses of microwave imaging while highlighting its benefits over traditional imaging modalities [15]. The intended antenna operates with an impedance matching of more than 10 dB between 1.47 and 4.7 GHz. At 4.3 GHz, the antenna has a peak gain of 3.8 dBi. To assess the antenna's impact on the human head, the Specific Absorption rate (SAR) of the intended antenna is examined. To accomplish this, a seven-layered head phantom is created separately [16]. A wideband antenna for ultra-wideband microwave imaging applications is presented in this study. Four star-shaped parasitic components, a rectangular slotted patch, and a tapering slot ground make up the antenna. It is demonstrated that the additional slotted patch effectively increases the gain and bandwidth. The suggested antenna design offers a broad impedance bandwidth ($S_{11} < -10$ dB) of 6.3 GHz (from 3.8 to 10.1 GHz), an accomplished gain of 6 dBi, and an efficiency of about 80% on the radiation bandwidth [17]. According to the simulation results, when the antenna is positioned straight on the skin of the breast as opposed to being put away from the skin, the current density in the breast skin is significantly lower and the current density in the tumor is significantly advanced. Although the antenna is presented in the context of microwave breast imaging in this study, there are many biological uses for the idea of installing an antenna directly on human skin [18]. After measuring the antenna's properties in an anechoic laboratory, it was determined that it had an omnidirectional radiation pattern and a working frequency range of 2.7–11.4 GHz, which encompasses the UWB frequencies and allows the antenna to be used for medical imaging applications. Lastly, the electromagnetic simulator was used to analyze the behavior of four of these antennas, which were positioned around a pragmatic breast model constructed of biocompatible materials. The findings were positive and showed the value of the devised antenna in the suggested application [19]. The suggested antenna uses partial ground planes with elliptical and circular radiators to create an ultra-wideband, low-profile construction that is optimized. The proposed ultra-wideband antenna (UWB) is used to detect malignant skin tumors using time-domain and SAR analysis. A rectangular human skin model comprising layers of skin, fat, muscle, and bone is made using the full-wave simulator [20]. A unique small directional ultra-wideband (UWB) antenna that can broadcast and receive signals in a specified direction is presented for use in skin cancer detection. The suggested antenna is made up of a partial ground, a T-shaped radiating patch with a microstrip feed line. The larger, unsteady bandwidth of the planned antenna is around 8.2–19 GHz. With its small dimensions and improved radiation patterns, the suggested antenna is suitable for usage as an array of antennas [21].

This chapter presents a novel technique where the antenna is designed at a 2.4 GHz operating frequency. The skin layers are designed and placed near the antenna to analyze RF characteristics. The gain, VSWR, and radiation pattern characteristics of the antenna are analyzed by simulating the antenna structure using HFSS software. The S parameters are absorbed in the presence and absence of tumor and are compared, as shown in Table 13.1.

## 13.2 ANTENNA DESIGN

The designed antenna consists of a radiating patch of dimensions 29.7 mm × 101.7 mm. The substrate has a ground plane below it, and the radiating patch is positioned above it. The substrate is of dimensions of 49.6 mm × 123.6 mm and a height of 3.6 mm. Both the substrate and ground plane have the same width and length. The typical range of dielectric constant for substrates in antenna design ranges from 2.2 to 12. With a dielectric constant of 4.3, FR-4 epoxy is the dielectric material employed in this application. Both the feed line and the radiating patch are etched onto the substrate. The feed line is used to serve the patch with microwave input. The feed line is of length 15.5 mm which is at the same height as the radiating patch. The feed used for the proposed antenna design is an inset feed. Figure 13.1 shows the flowgraph of the design and simulation process of the antenna in HFSS.

## 13.3 DESIGN EQUATIONS

The dimensions of the proposed antenna can be calculated using the following equations.

1. Height of the patch:

$$h = \frac{0.0606\lambda}{\sqrt{\varepsilon_r}} \tag{13.1}$$

*Table 13.1* Comparison of designed antennas for cancer detection

| *Name of substrate material* | *Dielectric constant of substrate material* | *Patch shape* | *Resonant frequency (GHz)* | *Gain (dB)* | $S_{11}$ *(dB)* | *VSWR* | *Reference* |
|---|---|---|---|---|---|---|---|
| FR4 PCB | 4.3 | Quasi-rhomboid shaped bow-tie | 4–8 | – | < −10 | – | [2] |
| FR4 | 4.3 | Vertical slot H–shape | 10.846–16.452 | 6.36 | −23.138 | 1.11 | [4] |
| Silk | 2.65 | Z-shaped slot | 2 | 1.19 | −12.5 | 1.36 | [5] |
| PTFE | 2.02 | Elliptical | 2.4 | 7.3 | – | – | [7] |
| Polyamide | 3.5 | Circular | 2.4 | 1.12 | −10 | – | [9] |
| FR4 | 4.3 | Planar inverted F | 5 | – | −20 | 1.21 | [10] |
| FR4 | 4.3 | Star | 11.72 | 10.9 | −23 | 1.17 | [11] |
| FR4 | 4.4 | Circular | 2.45 | 1.89 | −19 | 1.18 | [13] |
| Rogers RT | 2.2 | Tapered slot, Star | 3.88 | 6 | −18 | – | [17] |
| FR4 | 4.4 | Two rectangular steps |  | 1.3 | −33 | – | [19] |
| FR4 | 4.3 | Elliptical | 3.6 & 7.8 | – | −25 | – | [20] |
| FR4 | 4.4 | T-shape | 8.2–19 | 7 | −21 | 1.2 | [21] |

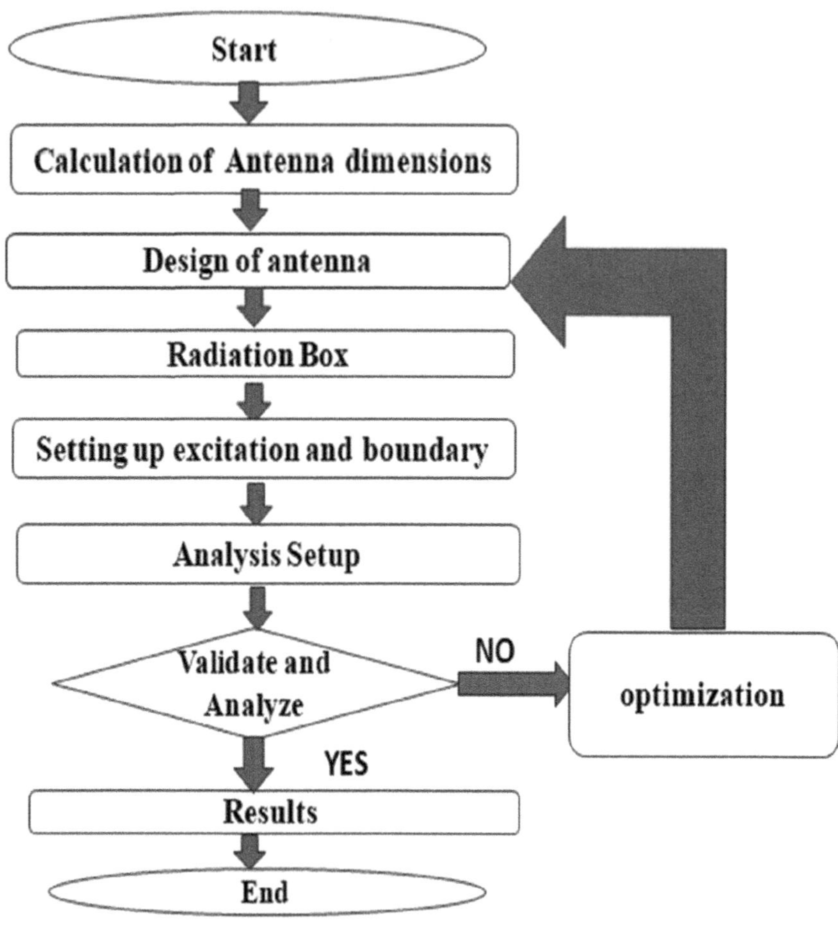


*Figure 13.1* Flow graph of design and simulation process of antenna.

2. Length of the ground:

$$L_g = L + 6h \tag{13.2}$$

3. Width of the ground:

$$W_g = W + 6h \tag{13.3}$$

4. Length of the patch:

$$L = \frac{c}{2f_r\sqrt{\varepsilon_r}} - 2\Delta L \tag{13.4}$$

*Table 13.2* Dimensions of proposed antenna

| *Parameters* | W | L | H | $W_g$ | $L_g$ | $F_l$ |
|---|---|---|---|---|---|---|
| Unit (mm) | 101.7 | 27.7 | 3.6 | 123.6 | 49.6 | 15.5 |

5. Total length:

$$L = L_{\text{eff}} - 2\Delta L \tag{13.5}$$

6. Effective length:

$$L_{\text{eff}} = \frac{c}{2f\sqrt{\varepsilon_{r\text{eff}}}} \tag{13.6}$$

7. Fringing factor:

$$\Delta L = 0.412h\frac{(\varepsilon_e + 0.3)(W / h + 0.264)}{(\varepsilon_e - 0.258)(W / h + 0.8)} \tag{13.7}$$

8. Effective dielectric constant:

$$\varepsilon_e = \frac{\varepsilon_{r+1}}{2} + \frac{\varepsilon_{r-1}}{2}\left[\frac{1}{\sqrt{1 + \frac{12h}{w}}}\right] \tag{13.8}$$

9. Length of the feedline:

$$L_f = \frac{\lambda_g}{4} \tag{13.9}$$

## 13.4 DESIGN OF SKIN LAYER IN HFSS

The skin possesses dielectric properties and healthy human skin has a dielectric constant of 38, and a tumor cell has a dielectric constant of 50. The model of a skin layer affected by a tumor has been designed using the above-mentioned values for material properties. A skin layer affected by a tumor has been designed [7] with an area of 49.6 where the human skin thickness varies from 0.4 to 4 mm. The designed skin layer has an average thickness of *2 mm*. For radiation analysis, the skin layer is positioned 36 mm from the antenna. The designed antenna is shown in Figure 13.2, and the skin layer in the presence of the antenna is represented in Figure 13.4. Figure 13.3 displays the skin layer of the human body designed in HFSS which consists of four layers such as bone, muscle, fat, and skin.

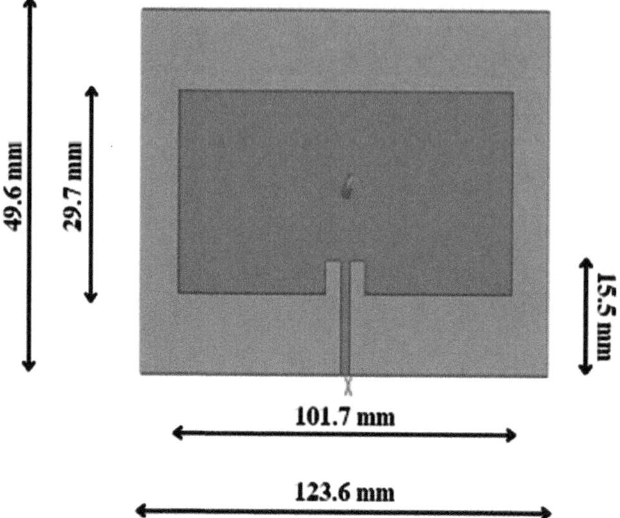


*Figure 13.2* Layout design of antenna.

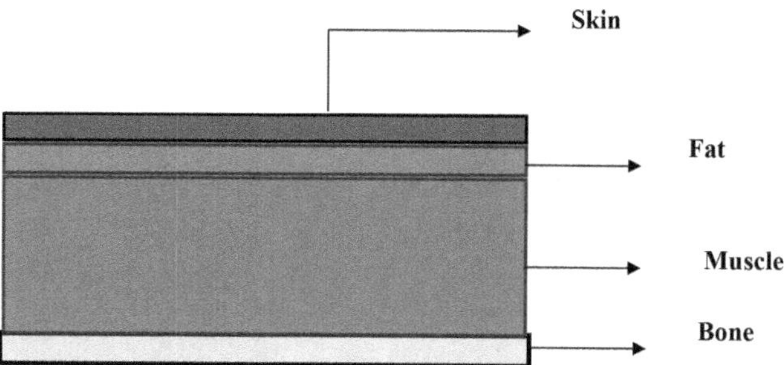


*Figure 13.3* Model of skin layer.

## 13.5 RESULTS AND DISCUSSIONS

The S-parameters such as Return loss and VSWR were analyzed for the designed antenna. The S parameters are used to relate the input port and output port parameters of the designed antenna. The amount of power transmitted from an antenna and reflected to it is represented by $S_{11}$, or return loss of an antenna. Nothing is radiated when the return loss is 0. The power transmitted to the antenna grows as the return loss falls.

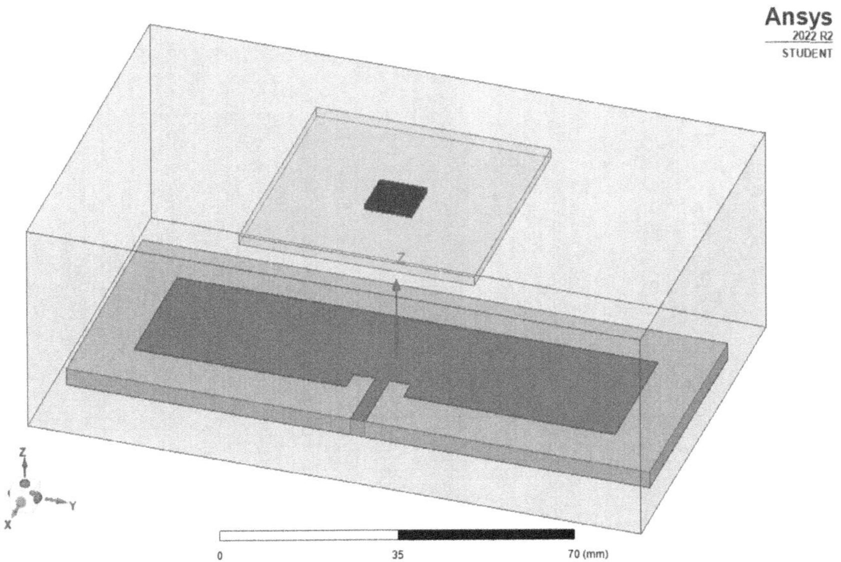


*Figure 13.4* Layout of the antenna in the presence of skin layers.

### 13.5.1 Return loss

The measurement of a reflected wave that travels or returns to a transmitter from an antenna is called return loss. The return loss for the proposed model is computed for the operating frequency of 2.4 GHz. For the designed antenna, the return loss plot is calculated in two scenarios. The return loss of –20.1075 dB is obtained under the assumption of healthy skin or the absence of a tumor cell, as displayed in Figure 13.5a. The return loss of –16.5139 obtained for a human skin tumor is displayed in Figure 13.5b.

### 13.5.2 VSWR

The voltage standing wave ratio is defined as the ratio of a standing wave's peak amplitude to its minimum amplitude. An ideal antenna has a VSWR value of 1 which reflects that the power fed to the antenna is received perfectly at the patch through the feed line via port. The VSWR plot for the designed antenna is computed for two different cases. The VSWR of 2.6146 is obtained for the case of healthy skin i.e., in the presence of a tumor cell which is represented in Figure 13.6a. In Figure 13.6b, the VSWR of 1.7985 for a human skin without a tumor is displayed.

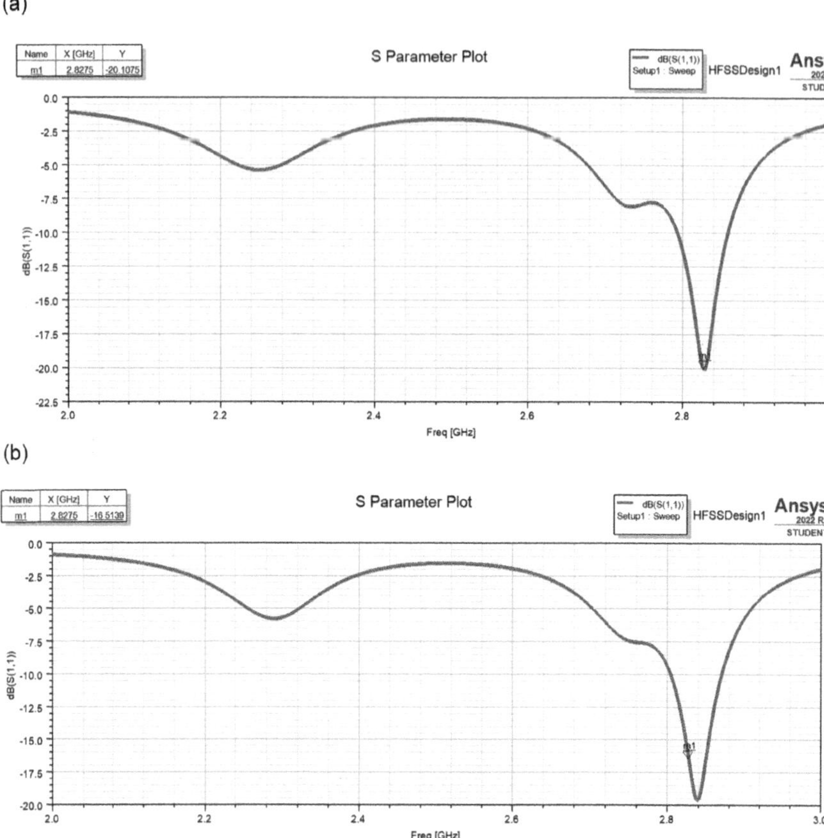


*Figure 13.5* (a) Return loss of antenna in the absence of tumor. (b) Return loss of antenna in the presence of a tumor.

### 13.5.3 Radiation pattern

The radiation pattern for the intended antenna which depicts the radiation properties of the antenna as functions of spherical co-ordinates at a phi angle of 0 and 90 representing the E-field and H-field is displayed in Figure 13.7.

### 13.5.4 Gain plot

The gain is the ability of the antenna to radiate its maximum power given as input by the transmitter in the desired direction which is measured in decibel (dB). The gain for the designed microstrip patch antenna operating at the frequency of 2.4 GHz is around 3 dB as displayed in Figure 13.8.

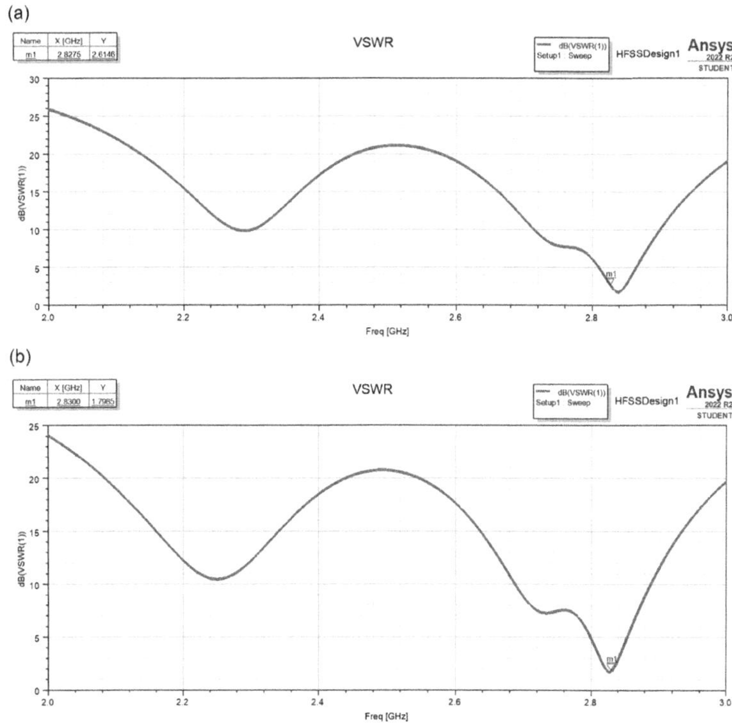


*Figure 13.6* (a) VSWR of the antenna in the presence of tumor. (b) VSWR of the antenna in absence of a tumor.

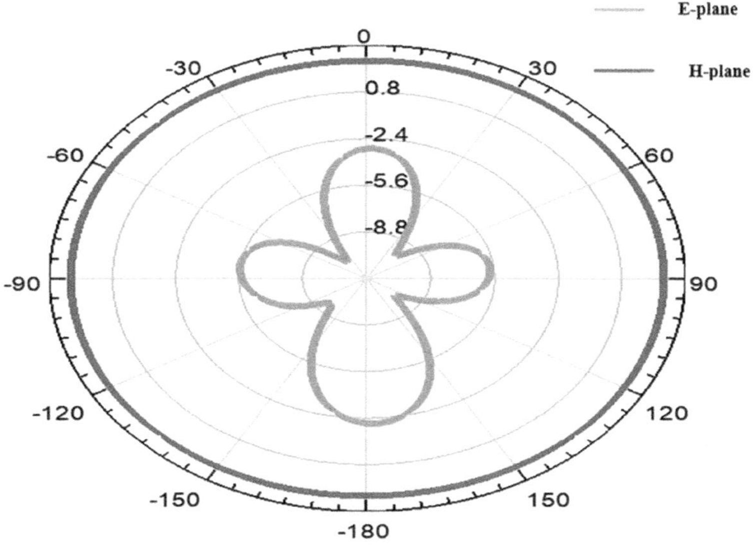


*Figure 13.7* Radiation pattern of designed antenna.

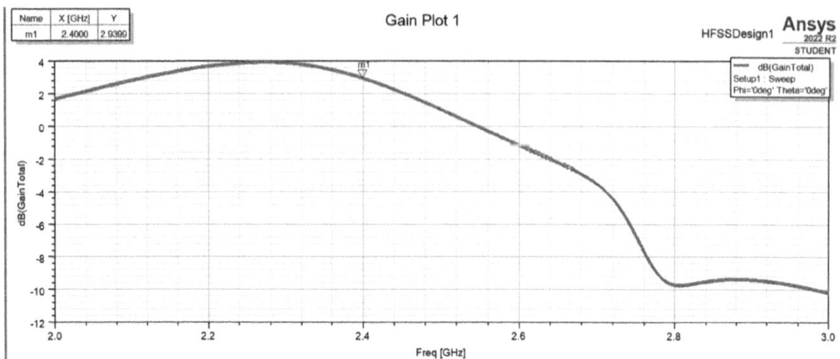


*Figure 13.8* Gain plot of the antenna.

*Table 13.3* Simulated results of the antenna in the absence and presence of skin tumor

| *Parameters* | *Antenna parameter with skin tumor* | *Antenna parameter without skin tumor* |
|---|---|---|
| Return loss | −20.10 dB | −16.51 dB |
| VSWR | 1.79 | 2.61 |

## 13.6 CONCLUSION

This chapter presents a technique to detect skin cancer by using an antenna. In order to detect skin cancer, an antenna is designed using HFSS software for the frequency of 1–5 GHz. The designed microstrip patch antenna with FR4 epoxy ($\varepsilon_{r=43}$, tand=0.0024, $h$=3.6 mm) as a substrate is simulated and the parametric analysis results of the antenna were obtained. The simulated VSWR for the designed antenna in the presence of a healthy skin cell and the VSWR of a skin layer with the presence of a tumor varies from 2.8 to 1.7 and the return loss of the antenna varies from –16 to –20 dB. The gain for the designed antenna which operates at a frequency of 2.4 GHz is around 3 dB. The radiation of the antenna has been analyzed. It is stable over the entire covered frequency band and at the two main planes E and H. Biological skin tissues have been developed in HFSS software by finding the dielectric property of the normal skin tissue. This tissue has been placed above the antenna with the air medium located at 36 mm apart.

## REFERENCES


1. Screening for breast cancer: US preventive services task force recommendation statement. *Annals of Internal Medicine,* 151(10), 716–727.
2. Chanchai T.; Supawat K.; Samran S.; Thanaset T. Breast cancer detection by using microwave ultra-wideband antenna. *International Journal of Electrical, Electronics and Data Communication.* 2018; 6(10), ISSN(p): 2320–2084, ISSN(e): 2321–2950.
3. Magthoom Fouzia Y.; Meena alias Jeyanthi K. Design of a novel microstrip patch antenna for microwave imaging systems. *International Journal of Engineering and Technical Research (IJETR).* 2014; 2(3), 119–123.
4. Jenisha J.; Madhan Kumar K. Design of H-shape microstrip patch antenna for wearable applications to detect the thyroid gland cancer cells. *ICTACT Journal on Microelectronics.* July 2020; 06(2), 928–933.
5. Rexiline Sheeba I.; Jayanthy T. Design and implementation of flexible wearable antenna on thyroid gland in the detection of cancer cells. *Biomedical Research.* 2018; 29(11), 2307–2312.
6. Rexiline Sheeba I.; Jayanthy T. Design and analysis of a flexible softwear antenna for tumor detection in skin and breast model. *Wireless Personal Communications.* April 2019; 2587, 175–185.
7. Sameer A.; Zahriladha Z.; Ahmad A. Miniaturized UWB elliptical patch antenna for skin cancer diagnosis imaging. *International Journal of Electrical and Computer Engineering (IJECE).* April 2020; 10(2), 1422–1429.
8. Gajanan Bhimrao T. Miniaturized UWB elliptical patch antenna for skin cancer diagnosis imaging. *International Journal for Scientific Research & Development.* 2021; 8(12), ISSN (online): 2321–0613.
9. Prakasam V.; Meghana R.; Sravani A.; Karishma Sk.; Divya B.; Chandu N. Tumor detection in skin using electromagnetic band gap structure antenna. *International Research Journal of Engineering and Technology (IRJET).* 2021.
10. Leeban Moses M.; Raju C.; Nair A.R.. Detection of skin cancer using PIFA antenna. *IOP Conf. Series: Materials Science and Engineering.* 2020; 2785, 180–188.
11. Amritjot Kaur N. Wearable antenna for skin cancer detection. *2nd International Conference on Next Generation Computing Technologies (NGCT-2016),* Dehradun, India, 14–16 October 2016.
12. Poorgholam-Khanjari S.; Zarrabi F.B. Dual-band THz antenna with SNG loads for biosensing and early skin cancer detection with Fano response: A numerical study. *Metamaterials: Design, Modelling, Simulation and Implementation.* 2021; 158, 1818–1900.
13. Top R.; Yasarunlu; Sinan Gultekin S.; Uzer D.. Microstrip antenna design with circular patch for skin cancer detection. *Advanced Electromagnetics.* March 2019; 18, 525–580.
14. Vignesh N.A.; RaviKumar R.; Rajarajan; S. Kanithan; E.S. Kumar; A.K. Panigrahy; S. Periyasamy. Silicon wearable body area antenna for speech-enhanced IoT and nanomedical applications. *Hindawi.* 23 May 2022; 25, 1512–1518.

15 Wang L. Microwave imaging and sensing techniques for breast cancer detection. *Micromachines*. 21 July 2023; 25, 1212–1218.

16 Asok A.O.; Nath S.J.G.; Vidhya S.J.; Kunju N. Brain tumor detection with compact monopole antennas using microwave medical imaging. *IEEE*. 24 Feb 2023; 12, 580–588.

17 Zerrad F.; Taouzari M.; Makroum E.M. Microwave imaging approach for breast cancer detection using a tappered slot antenna. *Materials*. 2023; 58, 1200–1208.

18 Shrestha S.; Agarwal M. Microstrip antennas for direct human skin placement for biomedical applications. *Progress in Electromagnetics Research Symposium*. 2010; 25, 250–258.

19 Martínez-Lozano A.; Blanco-Angulo C.; García-Martínez H.; Gutiérrez-Mazón R.; Torregrosa-Penalva G.; Ávila-Navarro E.; Sabater-Navarro J. UWB-Printed rectangular-based monopole antenna for biological tissue analysis. *Electronics*. 2021; 18, 580–588.

20 Rizwan S.; Kanaparthi V.; Kumar P.. An ultra wideband monopole antenna for skin cancer detection. *RAEEUCCI*. 2022; 25, 1200–1208.

21 Vanaja S.; Kalaivani S.; Manimegalai S.. Miniaturized directional ultra wideband antenna for skin cancer detection. *Journal of Social, Technological and Environmental Science*. 2017; 12, 1500–1508.

# Index

Note: **Bold** page numbers refer to tables and *italic* page numbers refer to figures.